Praise for *The Human Superorganism*

"Tremendously enjoyable.... *The Human Superorganism* really lays out the case for why the new research on the microbiome is a complete game-changer for how we view human health, and it offers this information in a comprehensive, readable, and thought-provoking manner. It informs how I approach my patients in my own practice."

—Susan S. Blum, MD, MPH, founder and director of Blum Center for Health and the author of *The Immune System Recovery Plan*

"A must-read if you are interested in disease, health, and medicine. Dr. Dietert has the unique ability to describe a new paradigm that is an easy read and understood at all levels of training or education."

—Gary R. Burleson, PhD, president and CEO of Burleson Research Technologies, Inc.

"In his startling and thought-provoking book, *The Human Superorganism*, Rodney Dietert shatters the conventional view of the human body by confronting the reality that most of the cells in our body are not our own. The book explains how an imbalance in the microbiotic ecosystem of our body has caused a sharp increase in allergies and other noncommunicable diseases in modern life, and it offers practical advice for fortifying and cohabiting productively with our single-celled partners."

—R. Douglas Fields, author of *Why We Snap* and *The Other Brain*

"In *The Human Superorganism*, Rodney Dietert challenges us to see ourselves anew, as stewards of our own personal ecosystems. By rejecting 'the new normal' of diabetes, obesity, cancer, and depression, we are empowered to learn how to feed our microbiome and begin healing ourselves from the inside out. The ultimate reward is a healthy internal environment that craves and is satisfied by what is truly good for us. In a world in which babies are born prepolluted with endocrine disruptors and other harmful chemicals, it may be our best hope of survival."

—Carol Kwiatkowski, executive director of the Endocrine Disruption Exchange and professor at University of Colorado Boulder's Department of Integrative Physiology

"Professor of Immunotoxicology at Cornell University, much-published superscientist Dietert wants us to look at humans as superorganisms—packed with microbes our ancestors have dwelled with comfortably but that we have sought to suppress. The result: a huge upsweep in noncommunicable diseases, from asthma to cancer to heart disease, now responsible for 63 percent of all human deaths. . . . Dietert tells us how we can radically readjust public health protocols—and our own."　　　　　　　　　　—*Library Journal*, Prepub Alert

THE
HUMAN
SUPERORGANISM

How the Microbiome
Is Revolutionizing the
Pursuit of a Healthy Life

RODNEY DIETERT, PhD

DUTTON
— est. 1852 —

DUTTON
→ est. 1852 ←

An imprint of Penguin Random House LLC
375 Hudson Street
New York, New York 10014

Diagram on page 126 by Spring Hoteling

LIBRARY OF CONGRESS CATALOGING-IN-PUBLICATION DATA
Names: Dietert, Rodney R., author.
Title: The human superorganism : how the microbiome is revolutionizing the pursuit of a healthy life / Rodney Dietert.
Description: New York, New York : Dutton, [2016] | Includes bibliographical references and index. | Description based on print version record and CIP data provided by publisher; resource not viewed.
Identifiers: LCCN 2015042307 (print) | LCCN 2015041110 (ebook) | ISBN 9781101983911 (ebook) | ISBN 9781101983904 (hardcover)
Subjects: | MESH: Microbiota. | Health.
Classification: LCC RA776.9 (print) | LCC RA776.9 (ebook) | NLM QW 4 | DDC 613—dc23
LC record available at http://lccn.loc.gov/2015042307

Printed in the United States of America
1 3 5 7 9 10 8 6 4 2

To my wife,
sweetheart, best friend, brilliant writer, and favorite novelist,
Janice

CONTENTS

CONTENTS

THE
HUMAN
SUPERORGANISM

INTRODUCTION:
THE NEW MEDICAL LANDSCAPE

The twentieth century was filled with ideas, discoveries, and inventions based on the benefits of freeing humans from bacterial, viral, and parasitic contamination. Wonderful scientific discoveries significantly reduced infant mortality, lengthened life spans, and drove medical technologies. However, the fundamental approach to human biology behind these advances unintentionally ushered in an epidemic of diseases afflicting humanity in the twenty-first century.

The two fatefully mistaken fundamental concepts were:

1. Humans are better off as pure organisms free of microbes.
2. The human (mammalian) genome is the most important biological factor in creating a better future for humans.

In the first section of this book I demonstrate how and why these two misguided principles came to underpin our understanding of medical science, producing a flawed paradigm that had untold adverse effects for the long-term health of our species. The desire for a biological purity that doesn't exist and the dream of a future for medicine built solely on the human mammalian genome have led us astray. The perspective I present flies in the face of much medical history and contradicts the thoughts of many brilliant Nobel Prize–winning scientists and educators. Nevertheless, our understanding of who and what humans truly are is undergoing a profound shift. It is a shift more and more researchers in the scientific community are recognizing.

In 1890 Robert Koch, a German physician and microbiologist, presented what later became known as Koch's postulates. These notions drove the infectious-disease paradigm of human medicine. They are simply four criteria used to establish the causal relationship between microbes and disease:

1. Every time a particular disease shows up, the same bug (bacteria, virus) thought to cause the disease must be present as well.
2. You must be able to remove a sample of the bug from the person with the disease and grow that bug in a lab.
3. You must be able to take the lab-grown bug and transfer it to a healthy animal or person and produce the same disease.
4. Then you must be able to take a sample of the bug from the animal or person you made sick and show that it is the exact same bug as the lab-grown sample.

Using Koch's criteria, specific microbes were rapidly found to cause many of the killer diseases of the early twentieth century, including typhoid fever, cholera, tuberculosis, and influenza. It soon became obvious that if you could kill the microbes causing disease and keep humans free of these pathogenic bacteria or, alternatively, produce protective immunity against some viruses using vaccines, you could reduce the burden of the killer infectious diseases.

And so we happily entered the era of antibiotics for bacterial diseases and vaccines for certain viral diseases. Penicillin was a game changer during World War II. Previously, wounded soldiers often died of subsequent bacterial infections. The only available drugs were quite toxic. But large-scale production of penicillin allowed field treatment of soldiers as well as treatment connected with surgeries. This helped to prevent death from gangrene as well as septicemia (blood poisoning). Some have called penicillin the greatest weapon developed during World War II. Ironically, it was a weapon against pathogens and thus a lifesaver.

Antibiotics helped to control cholera and typhoid fever. They mostly supplanted what had been the only strategy for dealing with deadly bacterial diseases: to separate and isolate patients until they died. Two diseases that killed countless victims and tore families apart were tuberculosis, also known as the White Death, and leprosy. Tuberculosis (TB) was known to the Egyptians and Greeks and killed an estimated one billion people during the nineteenth and twentieth centuries. For the first half of the twentieth century we sent TB patients to special hospitals called sanatoriums. Relative control of the disease was not possible until the development of a second generation of antibiotics (streptomycin) beyond penicillin. During the early twentieth century, TB sanatoriums dotted the US landscape and were essentially a place to keep patients comfortable as they were waiting to die. In 1919, there was even an unincorporated town established in Texas bearing the postmark "Sanatorium, Texas." Patients with leprosy, an ancient bacterial disfiguring disease, were sent to special isolated colonies. The leper colony on the island of Molokai is thought to have housed at least 8,000 individuals who were forcibly relocated between about the 1860s and 1960s. There they lived together until death. Antibiotic treatments saved lives and allowed families to stay together.

Viruses were every bit as feared, and this spurred the push for vaccines. Polio is a virus-induced illness that attacks the nerves in the spine, creating a debilitating, neuromuscular condition that can be fatal. Because children were more likely than adults to get the disease, it struck terror into the hearts of parents for several decades of the twentieth century. Though crippling to many, polio was responsible for only

6 percent of deaths in children who were five to nine years old in the early 1950s.

Perhaps the most prominent polio victim was President Franklin D. Roosevelt (FDR). Due to his own experiences and struggles with this disease, FDR became a medical philanthropist. This began with his trip to the mineral baths in Warm Springs, Georgia, to experience their healing properties. He was so impressed that he bought the site, created a foundation in 1927, and persuaded his law partner Basil O'Connor to run it. In 1933 FDR and O'Connor engaged in some early crowd funding when O'Connor began coordinating a Birthday Ball each January on FDR's birthday to raise money for polio-patient care. The balls were such a success that in 1938 they were merged into the national organization that eventually became the March of Dimes.

Most important, FDR's polio led him to initiate a major research effort to find a way to eradicate the disease. In 1954 the largest, most expensive medical experiment of that time was conducted. It involved using a killed-virus vaccine developed by the University of Pittsburgh's Dr. Jonas Salk. More than one million young children received either Salk's killed-virus vaccine or saline in a randomized double-blind study (neither the children nor their doctors knew which they were getting) that cost more than $5 million. When the study was completed, the National Foundation for Infantile Paralysis (NFIP) approved Salk's vaccine, and the specter of polio has rarely reared its head since.

It is human nature, at least in Western civilization, to identify a culprit when something goes wrong. We prefer to avoid considering how a complex biological system might be nudged toward better health. Maybe, unlike with polio, a single factor can't be used to achieve better health. This kind of problem is harder. But when it comes to human health, it is no longer avoidable.

Unknown in the golden age of the infectious-disease medical paradigm, these new treatments had a deadly side effect. Penicillin didn't only destroy the bacteria that sickened and killed so many in droves; it was nondiscriminatory as to which bacteria it killed. Unfortunately, it destroyed friendly bacteria right alongside deadly bacteria. The us-versus-them mentality viewed purging the microbes and creating a biologically pure human as the ideal outcome. This was the guiding path throughout the twentieth century in response to tuberculosis, typhoid fever, influenza, leprosy, and polio, and it has been hard to shake that war against microbes in the face of today's drug-resistant infections from HIV and mad cow disease to Ebola and MRSA (methicillin-resistant *Staphylococcus aureus*).

The idea was logical in the face of mortal epidemics and twentieth-century biology. But things have changed. We have met the microbes, and they are us.

The big problem with striving for human mammalian purity and spending much of the twentieth century obsessed with killing microbes is that it goes against our very nature. We are, as whole healthy humans, composed of thousands of microbial species and about 100 trillion cells. But the majority of those cells are microbial. If we indiscriminately wage war on microbes, we wage war on ourselves. For example, recent estimates of just bacterial cells range from a low of 57 percent to a high of about 90 percent of total human cells.

In humans there are more than 10,000 different microbial species in residence, although no single person carries them all. In one person with a healthy microbiome you are likely to find approximately 1,000 different gut bacterial species, with another 300 species in the mouth, 850 on the skin, and tens to hundreds in the urogenital tract. That is not counting the viruses, fungi, and parasites that also make up our microbiome. One square inch of our skin can contain up to six billion microorganisms, and you have about 3,000 square inches. You wear billions of microbes every day of your life. Different body locations vary widely in the specific microbes taking up residence. For example, if your feet have fewer bacterial species than your forearm, they more than make up for it given the fungi that absolutely love to live on your sweaty toes.

We are not just mammals. Not by a long shot. We need our microbial co-partner species. They have been there for centuries helping to support our ancestors. It is only recently that we have unintentionally cut them out of our lives in our modernized world of antibiotic-administered, formula-fed, cesarean-delivered babies growing up in urban environments, surrounded by hand sanitizers and antibacterial soaps. In doing so, we have compromised our own health. A new biology is emerging, demanding a different way of thinking about what it means to be human, to be whole, and to pursue a healthy life on earth for ourselves and our children.

In integrative medicine, practitioners talk about caring for and treating the whole human. It is a useful approach. The challenge for us now is that treating "the whole human" usually meant considering all the physiological, psychological, and spiritual systems together in approaching nutrition and medical treatment strategies. But now we must move far beyond those familiar notions to consider what is good for our microbes. The first revolutionary step in embracing the new biology is to start thinking of yourself as more than simply a mammal.

This flies in the face of some very basic biological principles you probably learned early on in grade school and that were first set out by the brilliant eighteenth-century Swedish biologist Carl Linnaeus. Linnaeus founded the field of taxonomy in which the identity and relatedness of biological organisms could be shown. He brought order to what seemed like biological chaos, and his work guided generations of evolutionary biologists and especially inspired Harvard's Stephen Jay Gould. While taxonomy remains vitally important even in analyzing the microbiome, the problem lies with the idea of species separation. The assumption has been we are a single species. Using the old taxonomy of Linnaeus, we would be categorized as *Homo sapiens*, a type of mammal, but operationally, that would be largely wrong. It would be wrong not just in how our bodies are composed but also in the genes we transmit to our next generation. Our operational taxonomy is that of a single species labeled as a human mammal. Instead, we each are a superorganism made up of thousands of species, biologically diverse. Be proud.

The New Face of Human Disease

Human diseases of the twenty-first century present new challenges. They are what we used to call chronic diseases and now refer to as non-communicable diseases, abbreviated as NCDs. They were originally called chronic diseases because they persist in the individual. Unlike a cold produced by a virus, these diseases do not go away in a week. In fact, once you get them, you often have them for life. We don't transfer them by sneezing or coughing, but they disable and kill just the same.

NCDs include allergies, cancer, heart disease, obesity, and even psychological disorders such as depression. They look nothing like what our ancestors encountered even a century ago. These twenty-first-century diseases seem to have emerged out of nowhere. They have changed not only when we die and what we die of, but also how we *live*, meaning quality of life, limitations, and challenges we face while we are alive. These new diseases comprise an epidemic, one we are as of yet mostly unprepared to deal with.

The present, growing epidemic is deadlier and costlier than influenza, measles, and Ebola combined. In fact, according to the World Health Organization, NCDs kill almost three times as many people (68 percent of deaths) as infectious diseases (23 percent of deaths). Yet NCDs tend to be a hidden epidemic. We have government organizations and academic departments in place to fight infectious diseases, but NCDs, as a whole, not so much. The efforts that do exist are usually partitioned into piecemeal programs directed at only one type of NCD such as cancer, obesity, heart disease, autism, or Alzheimer's disease. Comprehensive efforts to address NCDs have lagged way behind this epidemic.

The NCD epidemic is not restricted to any one culture, socioeconomic class, or geographic area. Almost three-quarters of the deaths due to NCDs occur in low- to middle-socioeconomic countries, although, proportionally, the rate is greater in more affluent countries, where NCDs cause up to 87 percent of all deaths. Distressingly, the epidemic only promises to get significantly worse in the years to come. But have you heard anything about this epidemic on the news—CNN, Fox News, or Huffington Post? Is it part of your Facebook feed or Yahoo Alerts? Is it trending yet on Twitter? No? If the epidemic is worldwide, then why not? Why the silence?

Unlike influenza, measles, and Ebola, the agent creating this epidemic is noncommunicable. You can't pass it to your family, friends, and neighbors by coughing, sneezing, or shaking hands. You can't see it spread. There is nothing to immunize against, and quarantines would be useless. Without the ability to prevent, vaccinate against, or cure these diseases, health practitioners are usually reduced to the helpless state of medically managing symptoms. In turn, this dramatically impacts personal productivity, quality of life, and socioeconomic viability. Individuals are reduced to a lifetime of drug management, which oftentimes creates a whole new layer of complications. Most drugs can have side effects, and as these side effects arise, they are often managed by prescriptions of yet more drugs. Our lives can become a series of alarms going off each day to remind us of the ever-increasing numbers of drugs we must take. Is this the life you planned for yourself? Is it what you want your children to experience?

The NCD epidemic is hard to pin down. We are used to chasing bacteria, viruses, or pathogens as a cause of human disease, and until recently many people were still trying to do that for NCDs such as cancer. But this is different. This epidemic involves an ecological system out of balance. Instead of being homogenous and relegated to one disease or one specific pathogen, such as a virus producing the flu, this epidemic comprises a myriad of different illnesses that each target different organs in the body and involve different medical treatments. Because of this, the epidemic has been more difficult to identify as a whole, harder to spot, and more challenging to nail down. It has been easier for health professionals and politicians alike to dismiss. Seeing it requires a new perspective, a new paradigm of human biology.

The new face of disease is there to see in our lives every day. It is anywhere people and the environment interact. They struggle to breathe the air, to eat the food of their parents, to move about, or in some cases to gather in crowds. They have to be increasingly cautious of their surrounding environment and how they interact with it. Those interactions are now dangerous for an ever-increasing percentage of people who are growing up sick, often isolated and seemingly ill matched for today's world through no fault of their own.

Welcome to the twenty-first-century epidemic of NCDs. In addition to causing 68 percent of all deaths, NCDs are the number one cause of disabilities, and a massive drain on our economies. In fact, it is estimated they will cost us $47 trillion per year in just a little more than a decade. They are already a global crisis requiring the highest level of attention of the World Health Organization and the United Nations, because the severity of NCDs appears to be increasing.

The NCDs are all too familiar ailments like autism and the autism spectrum disorders, food allergies, Alzheimer's, arthritis, asthma, cancer (all of them!), heart disease, celiac disease, diabetes types 1 and 2, inflammatory bowel disease, lupus, metabolic syndrome, osteoarthritis, sarcoidosis, thyroiditis (both Hashimoto's and Graves'), hypo- and hyperthyroidism, and the list goes on. From mental health, to what we can eat, to our very bones, this amazing and frightening spread of diseases targets just about every place in the body.

Beyond premature death, the NCD epidemic takes a toll on our everyday lives. How hard is it to cater to your six-year-old's birthday party guests? Mystified parents who feel they are doing everything right for their families are witnessing a loss of health and function in their children while the medical community is seemingly incapable of a timely, effective response. We have been moving toward an increasingly invalid society, with more and more children unable to experience life as their parents knew it and, in many cases, facing uncertain futures as adults. In the future, will we be able to associate with one another in ways most meaningful for human families, communities, and societies?

Before we attempt to understand the rise of noncommunicable diseases, let us look over the consequences of this raging epidemic. To keep things simple, let's look at just one aspect of modern life. Since John Denver wrote "Leaving on a Jet Plane" in 1966 and Peter, Paul and Mary made the song famous in 1969, air travel has burgeoned. It has become an essential part of our economy, many work lives, and often our leisure time. I remember my grandfather, who was a city councilman in Texas, taking the first commercial jet flight from San Antonio to Dallas's Love Field. This was a 287-mile trip that, as a child, I recalled as being a long, hot, grueling drive taking hours and hours. Yet my grandfather made the flight to Dallas in just forty-five minutes, had photos taken, then returned immediately. The entire trip took just ninety minutes of flying time. I was so incredulous that this was possible, I kept asking my granddad, "You were really there?" Jet travel changed everything for long-distance travel. But with the rise of NCDs, air travel is changing again, and not for the better.

In August 2014, blond-haired four-year-old Fae Platten from Essex, England, boarded a plane with her parents on their way home from Tenerife in the Canary Islands. She had a severe peanut allergy, which her mother had alerted the airline to, and flight attendants announced three separate times that no one should open peanuts during the flight. At 30,000 feet a man four rows away opened a packet of peanuts and disaster struck. Fae's mouth immediately swelled; her lips blistered; she struggled to breathe and finally passed out. Only an injection of adrenaline from an EpiPen saved her life. All of this horror due to small particles of peanut dust recycled through the plane's air-conditioning system.

News reports called the man who opened the peanuts "incredibly self-ish." But was he? He was no more absentminded than any of us might be from time to time. Maybe it was something else. His misfortune was to be living during the time of the noncommunicable disease plague.

For someone with diabetes, a severe drop in blood sugar can be as life-threatening as a peanut allergy to little Fae Platten. Let's suppose the man had insulin-dependent, type 2 diabetes. The Mayo Clinic has a long list of guidelines for diabetics when they travel, particularly out of their country of origin.

First, diabetics are encouraged to get a supply of insulin for the entire time they expect to be out of their home country and a doctor's letter to go with it. The insulin must be the exact same brand and type that they have regularly been using, since any changes may cause alterations in their blood sugar levels. The insulin must also be kept in a cooled container. Not only that, the diabetic must take into account changing time zones, changes in altitude, and changes in diet. The diabetic person must test his blood sugar levels more frequently for unusual changes and adjust accordingly. A significant drop in blood sugar can cause the diabetic to lose consciousness and, if he doesn't get sugar quickly, to pass into a coma and die.

On top of that, the Mayo Clinic recommends that diabetics keep food with them at all times. One of the leading foods listed is peanut butter because it is an ideal source to both raise and stabilize blood sugar levels.

Now I have to wonder. Is it possible that the man four rows from Fae Platten was an insulin-dependent diabetic? Perhaps he followed medical guidelines and tested his blood sugar levels when the plane reached the cruising altitude of 30,000 feet. Was it possible that he noticed his blood sugar levels were plummeting and now had his own medical emergency on his hands? What a dilemma. To open the packet of peanuts could jeopardize a little girl's life. To not open the packet of peanuts could jeopardize his own life. Was this the scenario that day on the plane? Probably not, but it is a very real possibility. In fact, the likelihood of this exact scenario is increasing every day.

But a diabetic on a plane needing a blood sugar boost is not the only problem. What if the person had celiac disease? Another dilemma. The only free snacks ever provided on flights are peanuts, pretzels, and cookies. However, for the person with celiac disease, even the leftover traces of wheat on a baking pan can cause a severe reaction. So for that person, pretzels and cookies are out, and the peanuts are the only snack that is safe.

My wife and I were visiting Texas. We and an acquaintance went to a craft village. While there, we entered a small bakery where a woman was selling fudge. The treat looked wonderful to all of us; however, my acquaintance has celiac disease, so he used caution and quizzed the woman about the ingredients she had used in making the fudge. She brought out the box the mix had come in. We all scrutinized the ingredient list carefully; no wheat or gluten was in sight. Relieved and happy, we purchased several flavors of fudge and indulged in a few squares on our drive back to his house. By the time he got home, his gut was in agony, and he spent the rest of the night locked away in the bathroom. The next morning he returned to the craft village to query the fudge lady further. Come to find out, she had laid the fudge out on trays that had previously been used to bake cookies. Though they had been washed, enough of a trace of gluten had remained to make him violently ill for hours.

These are three NCDs that could easily all be present in people on the same plane, or the same cruise ship, or even in the same school cafeteria. The issue has become so prevalent and so challenging that a court settlement ruling was handed down in 2009 that now brings celiac disease and food allergies under the umbrella of the Americans with Disabilities Act, thus requiring accommodations be made particularly in schools and colleges. And those are just the challenges of dealing with food allergies, diabetes, and celiac. What about NCDs like autism or inhalant allergies? As recently as 2002, the prevalence for autism spectrum disorders (ASD) was 1 in 150 US children. By 2010, it was 1 in 68. In 2014, it was 1 in 45.

Recently, an Oregon family took their fifteen-year-old girl with autism to Walt Disney World in Florida. Individuals with autism are frequently very sensitive to temperature and the way things feel in and on their body. In the case of this girl, she absolutely needed to eat her food steaming hot. She also needed to eat very soon after getting hungry; otherwise she would have a meltdown, become agitated, and scratch due to the discomfort and the frustration from her limited ability to communicate. The girl caused some disruption on the plane as a result of this condition. The mother convinced the flight attendant to heat some food; the girl ate it, calmed down, and began quietly watching a movie. Still, the plane made an emergency landing in Salt Lake City, police entered the plane, and the entire family was escorted off the plane because of the earlier disruption. While not life-threatening, this particular NCD created issues for the flight staff and tremendous embarrassment, humiliation, and outright fear for the family. If the mother holds to her word, it will also mean a hefty lawsuit for the airline involved.

Unbeknownst to most people, obesity is also an NCD, not the result of a lack of willpower. It has also become an epidemic in its own right, having more than doubled in prevalence since 1976. At present, more than one-third of the US population is considered obese, with that figure estimated to rise to 42 percent by 2030. The implications for air travel are not trivial.

Given that airlines have resorted to packing more coach seats into the same amount of space, obese passengers are facing increasing challenges when flying. In the case of a 518-pound man from Wales, an airline forced him to book two seats for a round-trip flight to Ireland. However, the airline staff themselves were clueless about the policy requiring him to purchase two tickets and gave him two nonadjacent seats—one on the aisle and one on the window, with a seat in between—for the first flight. On the return flight, airline staff booked him in seats that were two rows apart and obviously useless to him. That may sound merely silly, but it is not an isolated case of obesity causing problems we prefer to ignore.

Kevin Chenais, a twenty-two-year-old, 500-pound Frenchman with a serious hormone disorder, flew to the United States from France for medical treatment. In 2013, a year and a half later, he attempted to return home on British Airways, which informed him that they "were unable to safely accommodate the customer on any of [their] aircraft." Kevin and his family who had accompanied him were forced to take a train to New York City and cross the Atlantic Ocean by boat in order to get home.

These kinds of responses are quickly becoming standard policies with air travel. Three US airlines, Southwest, American, and United, require passengers who are too large to fasten a seat belt to purchase two seats. And Samoa Air has implemented a policy of pricing air tickets according to the weight of the passenger.

While barely noticing, we are becoming a society so biologically dysfunctional that our movements near and far are becoming restricted. The 2008 Disney-Pixar animated film *Wall-E* featured a frightening futuristic view of superobese people. The movie implied that the obesity was caused by poor eating choices and lack of exercise. There is another explanation for this social phenomenon.

Beyond air travel, obesity is becoming a political and legal issue. In the spring of 2015, Puerto Rico introduced legislation that would fine parents of obese children and register them as child abusers. Granted, there are parents who don't or can't provide healthy food choices for their children. It isn't hard to find a young mother feeding her baby pieces of French fries in a fast-food restaurant. With a busy, challenging work schedule, it's far easier to let the television, DVD player, and Xbox become the babysitter—especially now that parents are being cited for "free-range parenting" in which children are allowed to play outdoors without adult supervision.

That Puerto Rican legislation ignores recent scientific evidence suggesting that childhood obesity can result from a dysfunctional microbiome. Parents who obediently follow the recommendations of physicians to have a cesarean section delivery (complete with prophylactic antibiotics), and often accept physician-prescribed rounds of antibiotics for their children's upper-respiratory and ear infections, unknowingly contribute to problems with their child's microbiome. These state-of-the-art medical protocols prevent the baby's microbiome from seeding and maturing as needed. This, in turn, significantly increases the risk of childhood obesity. So what is, in reality, a lack of understanding of human biology by physicians, parents, and politicians could soon result in a charge of child abuse.

This is the tip of the iceberg. There are hundreds of different NCDs, and each one comes with its own particular set of limitations and daily risks.

What if the very environment in which you live and work could kill you? Such is the challenge of Viscount Jan Simon, a deputy speaker in Parliament's House of Lords and a Labour Party member. He developed asthma that is triggered by severe allergies to perfumes, tobacco smoke, and chemical fumes. Just a whiff of perfume, aftershave, or cigarette smoke can leave him fighting for breath in twenty seconds. Due to this, he collapsed and needed oxygen after a baroness who had washed her hair with scented shampoo sat next to him. On another day, while serving as the deputy speaker, he was handed a message written on faintly scented paper. With one whiff, Lord Simon was gasping for breath and had to be helped out of the chamber.

By his own report, Lord Simon has not been to see a movie in a theater since 1986. He has been unable to travel by train, bus, or plane for as long. And a casual restaurant meal is a thing of the past. His wife has had to change every personal care product she has ever used. Her nose is so acute that she often goes ahead of him to detect scents that could cause an attack. Even visitors to his home must follow a strict list of instructions in order for him to safely entertain guests.

Already schools have peanut-free zones, parents must check menus before parties, and at-risk people in enclosed public transportation can do little to protect themselves from their fellow passengers. Is it possible that one day twenty years down the road we'll see the beginning of a new era of segregation having absolutely nothing to do with race and everything to do with life-threatening NCDs? Imagine schools with entire zones and facilities for people with different food allergies. Many teachers are already specially trained to administer emergency adrenaline for this new kind of disease.

Other societal adaptations are under way in response to how we have changed as humans. The US Centers for Disease Control and Prevention (CDC) first established its own scent-free workplace in 2009, encouraging others to follow. Additionally, the Centre for Occupational Health and Safety of Canada offers guidance for implementing a fragrance-free workplace policy. We are simply not able to tolerate things that our ancestors enjoyed. The issue of fragrance allergies is so pervasive that recently the European Union banned three particularly allergenic ingredients that are part of the formulation for many popular perfumes. Two major perfumes affected are Chanel No. 5 and Dior's Miss Dior. The former is credited as the world's most popular perfume (based on sales), dates to at least 1921, and was given a mid-twentieth-century popularity boost by Marilyn Monroe.

How could long-standing, favorite fragrances that were just fine a few decades ago suddenly become poisonous for more and more of the population? The perfume formulations didn't change; we did. There are more and more people like Lord Simon among us.

For my wife and me, the social restrictions of NCDs aren't just theoretical and based on lab research. We each have our own NCD challenges that very often include foods and odors. Recently, we both attended a New York conference where I was to give a keynote address and we were to give a daylong workshop together. The conference was intellectually stimulating, and the attendees were wonderful, gifted people. However, food was an immediate issue. What was available from the conference caterers managed to hit all of our food allergies and sensitivities. We were both ill the first night and next morning.

In stark contrast, I had just given an invited lecture at the International Scientific Association for Probiotics and Prebiotics (ISAPP) in Washington, DC. All of the attendees were polled for dietary restrictions ahead of time. And throughout the entire conference, a variety of food options were readily available. Of course this is the new normal—and a nightmare for any group organizing human gatherings. It affects everything from church functions to meetings of Masonic groups to school athletic awards ceremonies. The NCD epidemic seems to be leading us to a world of isolation from our peers, colleagues, and families. We are becoming a race of disabled people.

This book offers an alternative.

PART ONE

A SHIFT IN HOW WE THINK ABOUT BIOLOGY

1

THE END OF THE OLD BIOLOGY

What if the very basis of what a human is was radically different from what we were taught as children? On one hand it could be challenging. After all, the majority of my life is past, and I went through it merrily thinking I had humans pretty much figured out. But no. It turns out that I was wrong about the most basic and fundamental concept of what a human is. I am not alone.

Our instruction about what it means to be human usually begins in childhood. It covers both the inherent nature of humans and their biology. Mostly this instruction comes from our schools. But often it is also connected to church or religious gatherings. And of course, daily we get ideas about humans from family members. A parent or sibling might ask such probing questions as "Why would you do that?" or "What on earth were you thinking?" There were questions around my house presaging the idea of the superorganism such as "What part of you thought that was a good idea?"

Our communities and even government organizations may weigh in on the nature and/or biology of humans. School, church, and family presented me with rather complete ideas on what I was as a human being and how I fit into the world as we know it. My childhood teachings from these different sources did not always align precisely. Even the people I looked up to the most had different views. And that was fine. In my own case, none of the sources of information on humans insisted that my view had to be theirs. Well, with the exception of having a grasp of Darwin's view of evolution for biology exams in school—that was mostly required.

School presented me with an evolutionary view that humans represent the pinnacle of life on earth, honed from earlier life forms and proven to be biologically fit through a rigorous selection process. It was an extension of Charles Darwin's general view of biology and how species are challenged and change. That was the academic mantra I encountered throughout much of my education in biology.

I remember being drawn to an often overlooked book by Theodosius Dobzhansky, a very famous plant geneticist, an evolutionary biologist, and the scientist most credited with updating Darwin's ideas in the light of the discovery of genes in the twentieth century. While I appreciated his work in genetics and evolutionary biology, the book that attracted my interest was not one of his evolutionary biology tomes. Instead, it was a perspective on genetics and human nature titled *The Biological Basis of Human Freedom* (1954). In this book, Dobzhansky moved beyond the narrow study of genetics to tackle more holistic topics such as man's kinship with nature and the relationships between genes, environment, and culture. He espoused the benefit of cooperative behavior for the fitness and natural selection of humans. Of course, Dobzhansky was thinking about cooperation among humans. In this book I am thinking about cooperation within a human. We are the village. Dobzhansky's unique and broader worldview was part of what led me to major in genetics in college, and those research credentials in genetics eventually secured my faculty position at Cornell University. I admired the fact that Dobzhansky was thinking more broadly about the possibilities of how his science applied to humans as both individuals and as a society.

Near the end of my formal education a brilliant mind added a fascinating idea to Darwin's theory of evolution. Former Oxford University professor Richard Dawkins published *The Selfish Gene* in 1976. In it he proposed that human beings are essentially "gene machines" whose biological operation is determined as a result of their carefully selected human genes. It was a tight argument. It was based on a general twentieth-century understanding of mammalian biology, and that has turned out to be its flaw. If we are robots controlled by genes, what genes control us exactly? Some of your genes never switch on; some never switch off; some switch off for a while, then switch on again. Should we only count the ones that are on? But then who or what is doing the switching?

These insights from the field of epigenetics made the gene robot or selfish gene idea a bit outdated. But perhaps even more important, we now know that 99 percent of the genetic information within the space we call "you" is not from your genome. Your genes only account for 1 percent of what is guiding cells in and on your body. The problem is that every time we think we know what is going on in biology, someone discovers something that we are missing, and sometimes it is something really big. To rethink the selfish gene idea, it might be more accurate to say, if we were built as anything, we are microbial storage machines designed to pass our microbes along to future generations.

When I was growing up, the church presented me with a creationist view of humans as a heavenly designed, newly installed organism on the previously unpopulated earth. It also provided guidance about how humans should operate in the world. I picked the parts that resonated with me, and became flexible in my consideration of exactly how life's creation might have occurred.

My far-from-atheist family was somewhere in the middle of the range of various scientific and religious concepts concerning humans. Fortunately, they gave me the space to work out how I would grapple with these different views while forming my own perspective about humans, other species, the earth, and the universe. By age twenty-five, I was rather stable in this merged Darwinian-church view, and that continued on for decades. I was even content that I had an operational view of humans and life on earth. Satisfied, I continued on with this understanding of humans until recently.

Almost no one understood the real biological secret about humans. The best minds—the pillars of biology, including Charles Darwin, Theodosius Dobzhansky, Richard Dawkins, James Watson, and Francis Crick—were missing a key piece of fundamental information about human biology. Namely, we are not what everyone thought we were, starting with our own DNA, our genome. We are far from alone even in our own bodies.

James Watson and Francis Crick discovered the nature of DNA in 1953, winning the Nobel Prize in Physiology or Medicine in 1962. Just fewer than fifty years after Watson and Crick's discovery, scientists had what was thought to be the magic key to humans: an almost complete sequence of the human genome. We would now know everything that was needed about human nature, health, and disease with the publication of a virtually complete sequence of all of our DNA and the genes within it.

Planning for the Human Genome Project began in the late 1980s, and three phases of five-year plans were launched to complete the analysis. The project itself was housed within the US National Institutes of Health (NIH) under the National Human Genome Research Institute funded through congressional appropriations with an estimated cumulative price tag of approximately $3 billion. It involved several US federal agencies and some twenty different major research institutions and organizations in North America, Europe, and Asia. Legions of researchers were involved in the painstaking work. One of the side benefits from the effort was the new technology for molecular analyses that emerged from this massive undertaking. The result changed biology, but not in the way researchers had anticipated.

Ironically, one of the greatest cooperative scientific accomplishments of our lifetime put an end to the old biology. In February 2001 the most significant results of the massive, global human research effort were announced in the journal *Nature*. This was the pinnacle of the long march of the twentieth century's genetic revolution in biology. It was a benchmark achievement for the Human Genome Project.

What caused one of the greatest human science projects to end an entire era of biological thinking? Quite simply, the results. The findings of the Human Genome Project were stunningly different than anticipated. Before the project started, our genome had been estimated to contain approximately 50,000 genes. Based on the science of the time, leading scientists believed our genome must drive protein synthesis, metabolism, and cell and tissue development. Control of the genome would be an almost magic elixir for curing disease. There was nothing inherently wrong with the premise except how much it underestimated the role of environment. The short story on the long march to unravel the human genome is that the results were underwhelming.

Instead of 50,000 genes, the human genome has approximately 22,000 genes—less than half of what was expected, and not nearly the numbers estimated to account for the remarkable complexity and diversity of human biological activities. In fact, we barely beat out the roundworm in gene-encoding proteins. The roundworm's genome has approximately 20,000 genes. Humans as the pinnacle of life on earth? Well, maybe. But if we are special, it is not because the number of our genes significantly exceeds those of other species.

So that 22,000 result isn't just a number; it has resounding implications. Our mammalian genes were supposed to rule all, in keeping with evolutionary biologist Richard Dawkins's ideas. And mastering our mammalian genes would surely cure human disease. At least that was the larger end goal for the Human Genome Project. But how can something as complex as a human be controlled, develop, survive, and even thrive with so few genes? The answer is we don't. We fail to thrive, get sick, and die with only those genes. Those genes are not designed to support humans in leading long, healthy lives. They are only a small portion of our life-support system. That is why the results of the Human Genome Project have ushered in the new biology. The take-home message is that our basic mammalian genome is only a very small part of what makes healthy children grow into healthy adults and produce more healthy children.

The project's underwhelming results forced us to look for what lies beyond the human genome. We are not self-contained, self-sustaining organisms as pure human mammals. We are intended to be more than that. The human being is a superorganism.

The twentieth century was a time of remarkable scientific accomplishment. In a sense, we became masters of our domain. And our domain was the whole world, plus space travel that took us beyond our world. In the introduction, I described the mid-twentieth-century changes that occurred surrounding infectious diseases. As a child of the '50s, I was fortunate enough to personally witness part of that transition as innovative science led to the conquering of many deadly infectious diseases. Salk's and Sabin's polio vaccines were just becoming widely available, and I got to witness the iron lung (used to keep polio patients alive) gradually disappear from our society's landscape. Scarlet fever and diphtheria were no longer threats for every child. Many from my baby boomer generation still carry an upper-arm scar from the smallpox vaccination that is missing from most people of more recent generations.

We were pretty sure we knew humans, knew the environment in which we lived, and knew how to control both to our benefit. We thought if antibiotics are good, more must be better. We could embrace our new-found capacities in both chemical manufacturing and industrial development as well as in food production and transport. It was all good. After all, we could now create a new type of artificial environment in which we controlled what chemicals were released into the environment, and we could isolate ourselves from microbes. We could radically change food availability, food types, and diets. We could seal ourselves away and be totally in charge.

While the intentions were well placed, our fundamental misunderstanding of what makes us completely human led us astray. We thought we didn't need to exist in our environment, but in fact, we can't live without enmeshing ourselves in that environment.

What we did during the twentieth century under the old biology was simply to trade one set of health challenges for another. We reduced the risk of death from infectious diseases. However, we increased the risk of lifelong disability and premature death due to NCDs.

In 2012, I was invited to contribute to a special issue of the physics journal *Entropy*. This issue was to address an important and novel subject: the measurement that could be taken that would best predict a healthy life rather than one filled with disease. What could you measure? What would you measure if anything and everything were open? What is the biological sign that foretells our health? I was certain I had the answer—after all, it was my life's work of the prior thirty-five years. I spent part of a day trying to start the paper. But the more I read over my argument, I had to admit, the more unconvincing it seemed. I was deflated. Was I having a bad writing day, or did I simply have the wrong answer?

I went to bed frustrated. In the middle of the night, I awoke from a dream. Parts of it eluded me. But what was left was a new concept of what constituted a healthy human. It was something that could be measured in a newborn. It was the microbiome.

The human microbiome is simply defined as the collection of microbes that live on and in us. You will see the term used to refer to both the microbial cells and their genes. Another name given to this collection of microbes sharing our body is microbiota. The microbiome is not in just one location in the body but is distributed throughout virtually every body part that has some exposure to the external environment. Some of the locations include our airways (nose, bronchi, lungs), gastrointestinal tract (mouth, throat, small intestine, large intestine, cecum), reproductive tract, and skin. Different locations have different resident microbes that love to live there and sometimes only there. They are a part of you just as they were part of your ancestors.

The idea I awoke with, that we need both the human mammal and a full complement of microbes for a baby to be healthy, became known as the completed self hypothesis, as described in a 2012 paper I wrote. My hypothesis states that, unlike what I learned as a child in school and church, human mammals are not really viable. As mammals devoid of microbial partners, we lack what we need to exist, let alone thrive. In the completed view of humans, we are each a conglomerate of thousands of different species. If we are missing key species, then we are not healthy. We are a hybrid: Each one of us is a superorganism.

The previous view of humans was the best of biology and theology that existed at the time. But now we know differently. We are a microcosm of earth's species. They virtually encase us, existing both inside and outside of us. We have organisms in our healthy bodies, such as archaea, that also live in the most extreme places on earth such as deep under the ice in Antarctic lakes. They are in the deepest oceans, the coldest and harshest conditions, and places where no sunlight can reach. We are connected to our environment in ways we never dreamed of.

Let us return for a moment to Richard Dawkins's selfish gene idea. Here is Dawkins's basic premise:

> We are survival machines—robot vehicles blindly pro-grammed to preserve the selfish molecules known as genes. This is a truth which still fills me with astonishment.
>
> —Richard Dawkins, *The Selfish Gene*

The basic assumption is that the human mammalian genome largely determines human existence and behavior. In a 1998 interview on the PBS show *Faith and Reason*, Dawkins also described how human or animal behavior serves the interest of the genes that developed the nervous system. It is a logical idea in Darwinian evolution but assumes that it is human mammalian genes supporting the human mammalian nervous system, which in turn causes behaviors in the interest of those same mammalian genes.

As part of the new biology, we now know that is not the case. Our microbial genes, or what is called our second genome, is now known to drive behavior that supports the bacterial genes and their propagation (e.g., food cravings). If there is a truly significant "selfish gene" in humans, it is probably microbial and not mammalian. That was not predicted by the old biology.

As you will read in later chapters, our microbial partners, the microbiome, significantly impact human behavior. So who is in charge? Exactly who is driving this bus? Did humans acquire microbes to enable them to build a better human, or did microbes design a better human as a new and improved vessel for their subsequent generations?

Maybe that is missing the point. Maybe neither is completely in charge. But we know thousands of species built today's human superorganism together, so the best conclusion is that there are predominantly cooperative, coordinated genes of our multispecies superorganism—not selfish genes. Humans without microbes are sick. Microbes without humans have no home.

This relationship is, in fact, even more intricate. Research has shown that there is gene sharing between our microbes and our mammalian self. We are intermixed as an organism even at the cellular-molecular level. Many of our present-day genes were not ours to begin with. They were donated by past microbial partners. You are not what you and I were taught. You are more than that. You are a reflective microcosm of the world in which you live.

Does this new biology I am touting actually do anything other than muddy the waters surrounding decades of debate about the nature of humans and the universe? It is my contention that it changes everything and largely in a good way. This change may happen only gradually, but it will inevitably impact how health care is delivered and how both humans and the environment will be protected. It will change our understanding of interpersonal communication. It may even result in changes in cultural and political climates.

But the first step is to embrace the highly useful part of you that you cannot see.

2

――

SUPERORGANISM ECOLOGY

The difference between the old biology of you as a single-species organism versus the new biology of you as a multispecies superorganism is a powerful network of ecological interactions. Ecology is the study of interactions between organisms and their habitats or surroundings. It can involve plants, animals, and microbes, including mixtures of all three. Sounds simple! Let's dig in.

While I was studying genetics, I had to pass a series of three-hour written exams, each going into depth on a different subject of biology. Ecology was one of those exams, and I had to prepare mightily for it since it was not a main focus of my courses or study. The university was blessed with several premier ecologists at the time. The exam question was simple enough: "What is a niche?" In discussions of evolution you often hear of how some organism prospered because it found a niche, perhaps somewhere where there was not so much competition for resources. I don't remember my specific answer. I do remember that, thirty-nine years ago, I had thirty pages' worth to say on the subject during the three hours. I remember waiting and waiting for the results from these exams, since they could determine whether I would be allowed to continue my PhD study. Ironically, the results never came. We were told that the two or more premier ecology professors who made up the question and were grading the exam were unable to agree on the answer. This was not a minor thing. In fact, the entire multiple-exam system for PhD qualification in that department was, at least temporarily, suspended.

There is a long tradition in science of new big ideas having to battle old big ideas. Sometimes the battle is loud; sometimes those new scientific ideas can emerge gradually and almost unseen. But perhaps most often, it is analogous to a prolonged first-pregnancy labor: uncomfortable, if not downright painful, and a bit scary. But the birthing of new scientific ideas can be a blessed event, just like the results of that physical birthing experience.

The argument for a different, more ecological understanding of humans was introduced nicely by David Relman in the opening sentences of an article titled "Microbiology: Learning About Who We Are," published in the journal *Nature*. Relman begins by noting that the "dawn of the twenty-first century has seen the emergence of a major theme in biomedical research: the molecular and genetic basis of what it is to be human. Surprisingly, it turns out that we owe much of our biology and our individuality to the microbes that live on and in our bodies—a realization that promises to radically alter the principles and practice of medicine, public health and basic science." Relman makes the case that microbes so affect our individuality that we cannot easily separate ourselves from their effects. Our biological identity and health are intertwined with that of our microbial partners.

Ecology is often represented in popular culture in ways that more resemble a soap opera or reality TV show, *When Animals Attack!*, but the less biting, more encompassing title *When Species Interact* fits more easily into my argument. Generally, when we think of species interacting, it is usually the type of interactions that are external to the organism. Things that come to mind are a cattle egret sitting on a cow, a koala up a eucalyptus tree, a bee's encounter with a flower, or even my dog and his obsession with the doves that he is certain invade his territory each morning. Just to extend definitions, ecological interactions describe the relationships among species when they share a community space, such as your body.

Different labeling is used to describe different types of interactions among species. For example, if two species interact and there is benefit to both, this is called mutualism. If one species benefits and the other is neutral about it or unaffected, that is termed commensalism. Finally, when one species benefits at the expense of the other species, that is parasitism. Within you, the superorganism, all of these types of ecological interactions occur almost daily. Properly managing our thousands of different species is called good health. Mostly it has been accomplished unconsciously, the whole ecosystem producing us in all our inimitable complexity according to the laws of nature.

Most people are familiar with parasites from giving heartworm medications to their dogs, having heard about the effects of tapeworms on food intake and digestion, or even the alternative therapy of deliberate exposure to hookworms (called helminthic therapy) as a way to shift certain immune reactions and reduce allergies. Less well-known perhaps is that malaria is produced by a parasite that inhabits red blood cells. That parasite is a microorganism called *Plasmodium*. Of the hundreds of species of *Plasmodium* that take up residence in various animals and plants, five have the capacity to infect humans and cause malaria. Parasitism is the simplest and least interesting form of species interaction—at least for the purposes of this book. It is true that parasites can provide some benefits while overall being pernicious, but there isn't a lot new to say about them. We knew we didn't want them in our bodies, and we still don't.

Two ecological concepts drive the bus for the completed self hypothesis. The first term, "commensalism," refers to eating at the same table, and that is precisely what our thousands of co-partner microbial species do. This term refers to the majority of our microbes that live on and in us but do not normally produce infections. In fact, you will see them called commensals or commensal bacteria. Those producing infections are termed pathogens or pathogenic bacteria. The original terminology for commensal bacteria was developed during the old biology. This is important because our relationship with most of these bacteria is not what was previously thought.

In the commensal relationship terminology of our old biology, our gut bacteria were seen to benefit from the association with us since we ingest food they can use. Previously, we were thought to be unaffected by or neutral to their presence within us (a very mammalian-centric view). The interactions may be complex, but we now know virtually every microbial part of our microbiome exerts some effect on either our mammalian self or on the other microbes that are present.

Many of these relationships are not commensal but are mutualistic, the second key concept. We benefit directly by the presence of our microbes. Take, for example, the case where certain bacteria digest otherwise indigestible sugars that are present in breast milk. This bacterial digestion and the production of food metabolites provide the baby with much-needed nutrients that it cannot otherwise get. The bacteria are also getting fed. So both the bacteria and the baby's growing mammalian cells benefit in a mutual exchange.

We, as hosts, receive benefit from the microbes through the maturing of our physiological systems. The newborn baby is essentially incomplete until the microbes take up residence and help that baby's development. Thus my notion of the completed self. This ecology plays out over one's lifetime.

Up until a few years ago, immunologists thought that the baby had all it needed for a well-oiled immune system at birth. That is certainly what I was taught during my immunogenetics training in graduate school. That idea came about because immunologists could count and label cells and see that all of the cells of the immune system seemed to be present at birth. But the fallacy in this thinking was the assumption that the presence of these cells meant they were well-balanced, fully mature, and functioning well together.

In reality, the numbers and markers available really told us little about what would happen when the immune system was actually challenged, such as by an infection. And that is where we as immunologists got it wrong under the old biology. If those immune cells do not encounter our microbial partners and "grow up" side by side with them in the baby, the immune system will produce dysfunctional responses at some point later in life. We are set up for immune-based dysfunction and disease if we are incomplete and lacking a full set of microbial partners. It is as simple as that.

Two well-studied ecological systems have been useful as models for how the ecology of humans should be approached. They are the tropical rain forests that circle the equatorial regions of earth and the coral reefs found in coastal areas of several continents. Learning from these examples can keep us from reinventing the wheel with our own ecology involving the microbiome.

Rain Forests

In the 2014 documentary movie about the microbiome titled *Microbirth*, I used the analogy of a forest to describe mammalian humans growing in partnership with their microbiome. The type of forest I had in mind was a complex, rich tropical rain forest. Like the areas within us that are populated by our microbiome, a healthy tropical rain forest flourishes with a mind-boggling diversity of life. Such rain forests are thought to cover only about 2 percent of the earth's total surface but contain more than 50 percent of earth's species. Besides being important for human well-being and the planet as a whole, they are a good model for what happens among species when things change.

A large group of scientists recently catalogued the tree species of the Amazonian rain forest. They found approximately 16,000 different species of trees. However, not every species was equally represented. According to the Nature Conservancy, a four-square-mile section of a tropical rain forest can contain up to 700 different species of trees, 400 different species of birds, and 150 different species of butterflies. But these numbers don't reflect the contribution of a rare species to the ecosystem. Rare microbes within us can provide absolutely vital functions.

Among all the rain forest birds and butterflies are some of the best sources of medicinal plants. This area of science studying indigenous cultures and their medicinal plants, ethnopharmacology, has become important enough to require its own scientific journals and multiple scientific societies. The types of drugs from these medicinal plants run the full gamut from anticancer agents to natural antimicrobials. A couple of examples are antimalarials (quinine) from the cinchona tree and antileukemia drugs from the rosy periwinkle.

During the 1990s, I was fortunate to work briefly alongside Cornell professor Tom Eisner as we were both senior fellows in the Cornell Center for the Environment. Eisner had been dubbed the father of chemical ecology and was a strong and effective advocate for chemical prospecting in the rain forests because he saw that it could both benefit humans with new drugs and simultaneously preserve biological diversity within tropical rain forests. He was not just a proponent of this strategy but actively worked with both corporations and conservation groups to make it happen. When it comes to the ecological protection of humans, my own thinking is much influenced by Eisner's sensibility.

Think of how the rain forest is divided into layers for a moment. The tallest trees provide the scaffolding for the forest canopy, which can reach a hundred feet, and their crowns receive the largest amount of sunlight and precipitation. They are also very efficient in photosynthesis. The canopy is rich with wildlife, including monkeys, sloths, parrots, macaws, and butterflies. In a healthy rain forest, the lower-level plants receive only filtered sunlight with much less direct precipitation and fewer strong wind gusts.

The understory plants usually live with higher humidity but cooler temperatures as they normally receive more shade. These levels tend to stay moist. Most of the understory plants would be recognized as house-plants such as the philodendron. Various tree snakes, the coatimundi, and fruit bats tend to hang out at this level.

Large mammals like anteaters, along with termites, giant earth-worms, scorpions, and ants, call the floor level of the forest home. De-composition of plant material is the main theme here. Fungi in the lower levels help to recycle nutrients to support plant and animal growth in the lower levels of the tropical rain forest.

It is beautiful to imagine this layered ecosystem. But of course in the modern world there has been much disruption. Deforestation has been connected to human practices such as logging, road construction that fragments the forest into smaller sections, and the conversion of forest into farmland. The sequence of events as a forest loses its biodiversity and the overall effects of deforestation provide us with a useful model for what we can expect if our own biodiversity declines.

In the forest, when the tall canopy-topping trees become thinned out too much, everything changes, not just for those trees but also for all the wildlife living in and under those trees. Forest clearing for agriculture is an obvious change since whole sections of the tropical forest can disap-pear almost overnight. Less obvious changes can happen when roads intersect the forests: More trees end up on the boundaries of the forest, where wind, different levels of exposure to sun, and more dramatic changes in water levels can affect the growth and sustainability of cer-tain species. Animals that use those trees for food and/or housing will have lost their means of sustenance and safety. The numbers and dy-namics will change unless they have highly useful alternatives.

The understory plants, living at lower levels beneath the canopy, will receive what amounts to a local climate change if the protective canopy is degraded. With environmental change and canopy thinning, more direct sunlight streams into the understory in the forest. That area be-comes hotter due to the increased sunlight and subsequent evaporation.

The plants and animals that live there will experience changes to their housing and food sources. It is a row of dominoes, each falling in turn.

Deforestation and changes in habitat affect both the numbers of representatives within each species and the species diversity as well. There is a domino effect of changing the habitats and species diversity where alterations in one group seem to go along with changes in others. Recently, this precise type of relationship was found in studies of plant and fungal species on the boundaries of rain forest/agricultural land areas by a multicontinent research team.

Coral Reefs

A second, equally useful example of ecology and the interaction among multiple species is the living coral reef. While coral reefs are often out of sight, the postcards aren't lying. They represent one of the world's true treasures. Coral reefs are not only rich locations for a disproportionately high percentage of marine life per square mile, but they are also protective barriers for coastlines and water-purifying mangrove forests. The three largest are the Great Barrier Reef off the coast of Australia, the reef off the coast of Belize, and a reef associated with the Florida Keys. Coral itself is a living animal, similar to a sea anemone, that has a soft body and grows very slowly. Its limestone skeleton base provides protection and support for the delicate body of the coral.

Coral lives in a symbiotic relationship with algae known as zooxanthellae. The algae provide oxygen to the coral and energy via photosynthesis. Additionally, the algae generate sugars, which the coral needs as nutrients in an otherwise nutrient-poor environment. The coral provides inorganic carbon in the form of carbon dioxide to the algae and acidifies the local environment, facilitating photosynthesis by the algae. The different colors associated with coral are more from the algae than the animal. Millions of species live within or around a coral reef, and their survival is interlinked with the vitality of the reef. Among the most familiar animals are the seahorse, lobster, various fish, sponges, sea slugs, eels, sea snakes, starfish, sea urchins, and clams.

Like humans, coral reefs also have bacterial and viral partners. Research into the complex interactions within the coral reefs provides a useful guide for understanding how humans can work with their microbial partners, as well as the risks involved in the degradation of our internal biodiversity. The coral reef was the original source for the term "holobiont," coined in 1992 by theoretical biologist David Mindell. A holobiont is a host organism and associated species that, as a group, serve as a unit of evolution. For the coral reef, the holobiont is all the species that participate in and are dependent upon the life of the reef.

Recently, the coral reef with its rich array of species has also been used to describe humans and their microbial partners. A human being is also a holobiont. Anything less than a fully staffed, human-microbial holobiont is a deficient organism, an incomplete self. Coral reefs can be damaged or degraded, and so can the human microbiome. The results are similar, predictable, and potentially tragic.

Coral is very sensitive to environmental changes and can be damaged by a mere touch from a snorkeling fin or boat anchor. Water pollution, infectious diseases, overfishing, fishing via dredging, tsunamis, storms, and climate changes can also affect the health of the reef. As with the tropical rain forest, the numerous species whose survival is linked with the reef interact in various ways, and effects on one can extend to others as well. Coral reefs need clear water so that their algae co-partners can get enough sunlight for photosynthesis. It is a delicate balance, teaching us about the ramifications of extremes within the ecosystem. Water pollution and increased silt associated with higher-density coastal cities and dredging can block the necessary sunlight. This increases the risk of disease and degradation of the coral reef.

Because the beneficial microalgae provide the coral's beautiful colors, there is an easy measure for reduced algae health; the corals undergo photo bleaching. They begin to lose their vibrant colors. You mainly see the white limestone base shining through the water. Worldwide, there has been increased coral bleaching for decades.

Another type of coral destruction bears striking similarities to the human obesity epidemic. It is related to the overgrowth of a type of algae that does not support the coral animals. This particular algae (known as macroalgae) can choke out a great deal of marine life. Excessive nutrients in the water due to pollution, combined with reduced feeding by fish on the large algae, can lead to overgrowth of the harmful, weedy, large algae. In many ways this parallels obesity-associated inflammation in humans. Too much nutrient intake produces changes in our own ecosystem, bringing in microbial partners that want those specific fattening nutrients and change the environment accordingly. When the overnutrition occurs and algae overgrowth is initiated, the resiliency of the coral reef declines. If the fish that are supported by the reef can't clear the large algae out fast enough, the reef declines.

Ironically, society recently has turned appropriate attention to the health of complex biological ecosystems like tropical rain forests and coral reefs. A better biological understanding of the risks involved with the destruction of these natural resources has permeated our thinking. Beneficial conditions that support these ecosystems, as well as harmful factors that contribute to their destruction, are better defined. The big question is: When will we apply the same level of concern and mobilize the same commitment to action toward the protection of our own human ecosystem?

Your Garden

As mentioned in Chapter 1, the microbiome is a collection of thousands of different species of bacteria, fungi, and viruses. They come from all three domains of life: the Eukaryota, the Archaea, and the Bacteria. Skin is thought to have approximately 1,000 different species of bacteria, with the phylum Actinobacteria the most widely represented.

The gut microbiome is composed primarily of two different phyla of bacteria: Bacteroidetes and Firmicutes (e.g., *Lactobacillus* found in yogurt). But a lot of the most interesting differences appear to be happening at the level of individual bacterial species, specific bacterial genes, and metabolic profiles. Like the skin, the gut is estimated to harbor approximately 1,000 different bacterial species. Beneath the species level, there is also considerable genetic variation. Gut microbial gene numbers across human populations are estimated at just fewer than ten million, but a majority of individuals share only a minority of those genes. Some bacteria species can have multiple strains each, with somewhat different copy numbers of genes and characteristics. Within gut bacteria species, some strains can vary by as much as a quarter of their genes.

Recently, a consortium of researchers presented a 3-D map of the microbiota inhabiting human skin in approximately 400 different body locations. The findings reveal the merits of an ecological approach to understanding body-location variation of inhabiting microbes. Different bacteria are prevalent on the face, back, and chest, where there are lots of oil-producing glands, than are found in the groin area, which has a local environment that is warmer with increased moisture.

The site-specific variation of microbes extends beyond the skin and is a general theme for the whole body. Body sites are different ecological regions, varying in such things as acidity, oxygen content, temperature, food availability, and moisture. These local environmental differences affect the mix of microbes that can thrive in a specific body site. For the microbes it is the difference between living in Miami, Florida, or Point Barrow, Alaska. Despite all being part of the same gastrointestinal tract, the mouth, large intestine, and small intestine differ in the profile of inhabiting microbes. If you were interested in oral health, the status of the oral microbiome would be likely to have more direct relevance than that of the small intestine or skin. Similarly, the mix of microbes in the mouth is wildly different from those inhabiting the vagina. The microbes want to be where they can have access to food and can grow and thrive, and in large part we want them to be in that specific location.

When relocated to the wrong body site, otherwise friendly microbes can cause problems. For example, gut bacteria getting into the body cavity is one of the fastest ways to induce septic shock—and death. So having each microbe in its own gated community works.

Your microbes in different body sites have an ancestral history of interacting with your mammalian cells in that specific location of your body. A cooperative synergy has been established between them across centuries. Each is tailored to match up well with the others. And, as we will cover in subsequent chapters, they have shared everything from metabolites, such as short-chain fatty acids (SCFAs), to genes.

One way to think of this is that you are cultivating a microbial garden. Each set of microbes has its own specific requirements for growth and function. Some like it hot, others prefer it cooler. Some like it acidic, others more alkaline. Some like light while others abhor it. Similarly, some want oxygen while others just want to be free of it.

The microbes you hear about the most are actually a very limited array of your total microbiome. You hear the names lactobacilli, bifidobacteria, Firmicutes, and Bacteroidetes at the forefront of probiotic discussions. But there are other microbes, and many of them aren't even bacteria.

The same types of ecological changes that affect tropical rain forests and coral reefs can also result in our own loss of biological integrity. Donna Beales, of Lowell General Hospital, recently termed this "biome depletion." Our own cells and tissues are affected when they are deprived of their microbial partners. You may be thinking that with the thousands of species contributing to our microbiome, what does it matter if we lose a few? We can afford it. But ecological studies tell us that maybe we can't.

There are two weak links in the maintenance of healthy ecosystems like the human superorganism. First, predominant species may have a particular set of maintenance requirements that must be met for them to survive and maintain their status at the top of the pecking order. Species in the greatest abundance will consume the most food and certainly contribute the majority of metabolites and waste products, thereby affecting the overall environment of the ecosystem. In a way, they determine what is left for the rest of the species. Shifts in food availability and other conditions can impact their prevalence and affect the habitats of most other species as a result.

A lot of research has gone into this group of most prevalent species in diverse ecosystems such as the tropical forest and our own gut. But again, the prevalence of certain species does not always reflect their importance to the ecosystem. In fact, in highly diverse ecosystems like the rain forest and our gut, skin, airways, and reproductive tract, the rare species actually perform critical functions, such as promoting useful immune system maturation. In turn, they may be the most vulnerable when it comes to damage to the ecosystem. These are often referred to as keystone species. In such ecosystems, a lack of redundancy for supporting critical functions performed by the rare species is likely to be the tipping point in system failure.

In the tropical rain forest, 55 percent of the tree species that are involved in critical functions have only a single representation per sample. Take that single representative away and the local area is missing that critical function. Thinking locally in the gut, on the skin, and in the airways is important. Different regions of the gut harbor distinctly different microbial species that are tailored to support the specific bodily functions of that section of the gastrointestinal system (e.g., microbes of the large and small intestines are very different). The weakest link in each subsection of the gut should be made a priority when protecting the human ecosystem's health. And there is at least one example from lab animals that illustrates why this is the case.

One comparatively rare specialized gut bacterium called *Akkerman-*

sia represents only 3 to 5 percent of gut bacteria. Yet, these bacteria play an important role in communicating with cells in the gut lining and regulating mucus production. The mucus layer is critical for keeping other bacteria at a healthy distance from our gut epithelial and immune cells. If the comparatively rare *Akkermansia* bacteria are damaged and their numbers reduced in the gut, as appears to occur under certain environmental conditions, the critical function of gut mucus-lining maintenance is lost. Reduced *Akkermansia* numbers are associated with a form of obesity-promoting inflammation.

Not surprisingly, the *Akkermansia* bacteria were a relatively under-studied type of gut microbiota until recently. Given their low profile, there was no inherent reason to suspect their importance. Yet the loss of gut *Akkermansia* now appears to be a tipping point for a host of inflammation-related diseases and conditions. Protecting the weakest, most critical link that promotes a healthy microbiome and effective human physiology is likely to become the highest priority.

A wealth of studies in rodents and other animals shows us what happens when the microbiome is degraded, damaged, or even lost. The storyline strikes me as a little similar to the classic Frank Capra movie *It's a Wonderful Life*. We have the information to look ahead and see what the future brings for living with a damaged microbiome. It is not pretty. It is not something we would want for ourselves or our children.

When we lose our microbial partners, the path toward effective development and function is altered, and not in a useful way. The evidence supporting this has been around for a while. In fact, a comprehensive review from 1971 about the effects in lab animals when the microbiome is absent foretells exactly what happens when we are a single human mammalian species. Without those microbes, we face a life of biological deficiencies, illness, and death.

Take, for example, the forty to fifty years of producing and studying two types of mice: gnotobiotic and germ-free mice. These are mice specifically maintained in bubble-like conditions, eating sterilized food and, at least initially, delivering their babies by cesarean section. Gnotobiotic mice are completely free of bacteria, including normal microbiota. Essentially, gnotobiotic mice have to be provided with special nutritional supplements in order to survive. This is because the gut bacteria make specific nutrients that are required for a healthy life but are not produced by the mammalian cells. These include the fat-soluble micronutrient vitamin K. Levels of bacterially produced vitamins and other metabolites are critical for survival. For example, if gnotobiotic mice are fed standard rodent chow, they become sick within three days and die. Thymidine deficiency is also present, as bacterially produced thymidine is not available. Rodents completely devoid of a microbiome that are not provided with special bacterial metabolite supplements cannot survive.

There are additional issues. The cecum, part of the intestine, normally represents 6 to 10 percent of body weight in a rodent but, when lacking all gut bacteria, can swell to 20 to 25 percent of total body weight, and these complications can produce death. The heart reduces in size, and blood flow and oxygen delivery are reduced along with it. The animals have decreased motor activity as well. The immune system is defective, as are immune responses.

It is worth looking at what happens in these mice when they are deprived of a microbiome or even provided with a partial microbiome. First of all, they have to be maintained under germ-free lab conditions. If they are exposed to normal conditions, they will die of infections. According to one of the suppliers of germ-free mice (Taconic), probiotic mixes need to be supplied just to keep the mice alive. Otherwise, they are vitamin K deficient since that vitamin is made by the microbiome. Interestingly, it has been known for some time that antibiotic treatments in humans can significantly reduce the levels of vitamin K as the gut bacteria are killed off.

Just as with the tropical rain forest and coral reef, there are consequences to degrading or damaging the human microbiome garden. You need that to be whole, and you need access to a sufficient diversity of microbial partners to have a healthy and prolonged life.

3

———

THE INVISIBLE HUMAN
SUPERORGANISM

Humans are for the most part microbial. Scientists have estimated that by cell count you have ten times more microbial cells than mammalian cells in your body. When geneticists compared microbial genes to mammalian genes, they found we are even more microbial genetically; while humans only have about 22,000 mammalian genes, we carry approximately ten million microbial genes. The totality of your microbial genes, including bacteria, viruses, fungi, and parasites, has been called your second genome.

This second genome is important beyond just the numerical comparison. To change the mammalian genome means changing the chromosomes in every cell of the body. A chromosome is a threadlike structure found in all living cells that is made up of nucleic acids and proteins and carries genes. A mammalian gene makeover would be a daunting task given the number of chromosomes and cells that would have to be changed. But while it is difficult to change mammalian genes, it's comparatively easy to change the microbial genes. Basically, all you have to do is change your body's microbial mix and you change your microbial genes. That fact provides a powerful new strategy for improving our health and well-being.

Another challenge for changing the mammalian genome is the fact that many mammalian genes work in groups. The attempt to change one gene can set off a chain reaction in other genes, resulting in some genes not being fully expressed. This may cause some functions to be altered or even go missing. Even if you got the mix right, that most likely would not correct most noncommunicable diseases. Things like asthma, diabetes, obesity, and autism seem likely to require changes in many genes and metabolic pathways involving the immune system as well as organs and tissues. In any case, the overriding limitation to mammalian gene therapy is that it only targets less than 1 percent of our total genes (first *and* second genome).

Think of the possibilities of targeting your microbial genes—99 percent of your total genes—by changing your gut microbes or your skin microbes. That is not science fiction. Researchers and clinicians have already accomplished what is called a "proof of concept." They can make these changes, and they have methods that have already worked. As I take you through this chapter on the human superorganism, I will be stressing our genes, both microbial and mammalian; how they have affected our history and our present status; and how they are likely to affect our future and our children's future.

The New Family Rules

Right there inside you and on you, you are the world, just as Michael Jackson and Lionel Richie and a host of other artists sang back in 1985. You are a microcosm of thousands of species from this world. You are not alone. You are more than you ever thought you were. What is in you is also shared with millions of people from places all over the world. People you have never met. Places you have never been. Yet you are related to them microbially. Your microbes are related to those people living continents away and decades before you.

In the past, we only thought about shared microbes in a very negative context since they most often led to infections that swept the globe. This was how plague, smallpox, typhoid, tuberculosis, and the polio virus became epidemics. But your resident microbes that don't normally cause disease and support your body's maturation and function have circled the globe as well. Most of us don't think about our relatedness to a visitor from a continent other than that of our ancestors. But in total gene composition, including your microbial genes, a relationship is almost inescapable.

The connection with family is strong and has persisted through the millennia. Blood relations have been the basis of communities, tribes, and clans in many ancestral cultures. Blood relatives might wear clothing or other adornments signifying their loyalty, be it a Scottish tartan, a pattern of beads, or a specifically styled tattoo. Heraldic symbols on shields and crests, sometimes even family mottoes, labeled families as they marched into battle. In times when even many nobles were illiterate, they would also use these symbols as their official mark on contracts and letters.

Kinship has been a basis of politics for a very long time and continues even today—you can see it from Kazakhstan to Kennebunkport. To most of us, such affiliation is just the way the world works.

Microbial heritage and loss was not something that our ancestors could see, like the blood red of hemoglobin, or even understand until recently. Yet as a growing number of biologists around the world are beginning to recognize, microbial loss probably matters more than kinship for the human legacy parents leave their children—and their children's children. Drawing on his research, New York University professor Martin Blaser recently suggested in his book *Missing Microbes* that we cannot afford the loss of microbial diversity that has been created, in part, by the overuse of antibiotics. The benefits of establishing and maintaining a healthy, family-based microbiome are clearly set out in his rigorously argued book.

Much like those ancestors who wore a Scottish tartan, we should wear our microbial colors with pride and strive to protect and preserve them. As we shift our focus from our first genome, consisting of mammalian genes, to our majority second genome, our microbial genes, our own perspective on ancestry and legacy is likely to shift as well.

The battle of the sexes at the center of various culture wars is likely to be upended by the idea of ourselves as superorganisms. This is how fundamental the new perspective is. Relationships among men and women, husbands and wives, appear in a new light. My wife and I jointly authored a history paper detailing the underappreciated role of women in the history of goldsmithing in Scotland. We found that women heavily influenced who got to apprentice to which silversmith, based on the women's family ties, and what wares the silversmiths produced. Then men just trained and produced what was required.

Throughout history there have been two predominant inheritance types—patrilineal, dominated by the father's family, and matrilineal, led by the mother's family. Whether newlyweds lived with the father's family, termed patrilocal, or with the mother's family, called matrilocal, also had implications. There were also rules determining who paid whom for the privilege of marrying, who lived where after marriage, and who inherited the family's wealth and property.

While these rules might have been more tribal, similar rules extended to rulers (e.g., kingship) and who decided who ruled (e.g., some Native American tribal chiefs were chosen by the women). While both male- and female-dominated societies have existed and do exist today, approximately 80 percent are patriarchal. Anthropologists argue that war may be a driving force behind this. Matrilineal descent, in which mother-daughter inheritance dominated, was often due to less certainty over paternity. Protecting the integrity of the family's mammalian genome in the bloodline became the driving force behind patrilineal lineage.

The rules of kinship and succession certainly made life interesting. Take a specific case told over and over in movies, TV shows, and even operas. Henry VIII, king of England, had a lot of wives.

One could argue that English Protestantism arose because Henry VIII was unable to encourage more of his Y-chromosome-bearing sperm to perform their duty with his queens' eggs. Yet given biological understanding in the 1500s, the women took the blame instead. Even if he had had an adult male heir, the real genetic inheritance would have been through his wife's microbial genome. Since Henry did not have a long-surviving male heir, his daughter Elizabeth was eventually crowned queen of England. The daughter of Henry's second wife, Anne Boleyn, Elizabeth was three when her mother was executed. Since the microbial genome is 99 percent and inherited from the mother, Elizabeth was more a queen in Anne's lineage than that of heir-obsessed Henry.

Anne Boleyn's reported craving for apples during the pregnancy probably helped to craft the eventual donated microbiome for Elizabeth. Two weeks before delivery she retired to a chamber that has been described as a cross between a chapel and a padded cell. Ironically, as the baby's delivery approached, only women were allowed into Anne's chamber. That would seem to suggest that any bystander microbes donated via skin-to-skin contact with Elizabeth were from ladies of the court and not from King Henry. At three P.M. on September 7, 1533, Elizabeth was born via natural delivery, and the baby's biology was completed by Anne Boleyn's donated microbiome. Observers noted that the baby Elizabeth got Henry's red hair and Anne's dark eyes. Of course, what they didn't realize at the time is that Elizabeth had far more genes from Anne's body than from Henry's when ascending the British throne.

If you actually tally up the genetic contributions of Henry versus Anne to Elizabeth using the facts that Elizabeth had about 99 percent microbial genes and only 1 percent mammalian genes, it turns out that Henry donated only 0.5 percent of Elizabeth's total genes, with Anne Boleyn providing 99.5 percent of the genes, minus a few microbial genes that came from birth attendants and wet nurses who might have breastfed Elizabeth. Whose ancestral baby bottom graced the English throne? Mostly Anne's.

Though Elizabeth had a lengthy, powerful rule, she never married nor produced a royal heir. Succession to her throne created significant contention. Her cousin Mary, Queen of Scots, vied for that honor, and their dramatic confrontations are legendary. Were that happening today, they would command a reality series in their own right. To keep Mary from gaining the British crown, Elizabeth had her imprisoned and eventually executed. In an ironic twist of fate, Mary's son, James, who became king of Scotland while yet a baby, became Elizabeth's successor. And Mary lived on and/or in James via the microbes she contributed for his gut during delivery and the skin microbes she exchanged with him when she held him. Most of the microbes and genes that passed from Anne Boleyn to her daughter, Elizabeth, or from Mary, Queen of Scots, to her son, James, had nothing to do with mammalian genetic kinship. The microbes carried far more biological information. Maybe it's time for a new opera about Henry and Anne Boleyn.

You don't have to go back multiple centuries to find societies in which the inheritance of power, money, and even fame were governed by the mammalian male line. It prevails in some cultures and societies today, and its pernicious influence may even be growing. It stems from what I call 1 percent thinking: the idea that a male heir passing chromosomes from generation to generation is the true test of a family's worth, the true bloodline. It has virtually nothing to do with biology.

A preference for sons occurred in several agriculturally based cultures spanning the globe because they have tended to earn more money. The dowry system, where families with daughters had to pay the groom's family for the right of the women to marry, also arguably led to a devaluing of women in general. In the twentieth century, this archaic tradition meshed with new technologies in an alarming way. During the 1980s, prenatal sex identification changed things, mostly in China but elsewhere in the world, too. If a baby's sex could be identified in utero and sex-specific abortions were possible, well, the outcome was awful but perhaps not surprising—population selection against women.

Female fetuses were aborted while male fetuses were carried to term. This of course does have a long-term biological consequence.

The view that the male offspring continues the family line is based on the pseudoscientific idea that a continuous line of males passes on the true family genetics. But again, chromosomes passed by males across generations only comprise less than 1 percent of the genes that are a part of us. And here is a twist: The 99 percent of microbial genes passed from generation to generation are largely inherited through the women in a family. Apparently, some cultures like the ancient Picts in Scotland got it just about right.

China, with the world's largest population, implemented a one-child-per-family program in 1979. With some variations in different provinces, couples could only have one child. Two were permitted if the first was a girl. The policy was intended to last one generation, but it persisted. Given the culturally ingrained preference for a male heir and the ramifications of limited family size, the outcome led to a critical overabundance of male children and a shortage of females. With a present excess of forty-one million bachelors, according to the Population Reference Bureau, that figure is expected to grow to fifty-five million by the year 2020. Given the desperateness of the situation, China relaxed its one-child policy in 2013.

In India the situation is no better. A 2013 *New York Times* article looked at the "man problem" in India. Its author concluded that the excess numbers of unmarried men had led to increased violence against women. And the sex-selection problem during pregnancy is quite extreme in certain regions of the country, particularly where dowries are still culturally required for girls. While laws have been passed both to discourage sex selection and to do away with dowries, enforcement has been problematic.

So many historic and present-day conflicts, wars, views of succession, examples of racism, and even sex selection of offspring have been based on what we now understand are biological half-truths—all given currency by the dominance of the idea that you are what others can see or peek at. Advertising in our glossy online culture reinforces this image-based approach to human evaluation. We love body images. But that mammalian body image is not the real you. Essentially, you aren't just a body; you are a superorganism. When you want to find your core, when you want to understand what is deep inside you, when you want to control your health and moods and interactions with others better, you must seek out the genetic 99 percent of you that is microbial.

Baby, Meet Your Microbiome

The seeding of the newborn's microbiome occurs largely at birth. Prenatally, the baby is exposed to some microbes, such as bacteria associated with the placenta, and this no doubt helps with prenatal immune maturation.

The placenta has a much smaller community of bacteria, including the phyla Firmicutes, Tenericutes, Proteobacteria, Bacteroidetes, and Fusobacteria. The microbiome of the placenta seems to most closely resemble that of the mouth. The diversity of the placenta's microbes seems to be related to the baby's prenatal development. In a recent study from Beijing, China, researchers found that the placental microbiomes associated with normal weight versus low birth weight in babies differed significantly. Lower-birth-weight babies had placentas that were comparatively barren in terms of bacterial diversity and were also reduced in the percentage of *Lactobacillus* bacteria.

Maternal environment, including diet, stress, and drugs (e.g., antibiotics), plays a large part in crafting the array of microbes that will seed the baby. The birth event itself is the single most important step in the seeding process. It is during vaginal delivery that the baby is exposed to both microbes in the vagina and those from the mother's cecum, a portion of the large intestine near the appendix. Bacteria that can grow with or without oxygen, such as Enterobacteriaceae bacteria, are among the first to appear in the newborn's gut, and these are replaced shortly thereafter with several different types of oxygen-hating bacteria *(Bifidobacterium, Bacteroides,* and *Clostridium)*.

These are the founding microbes that are the first co-partners of the newborn. These beautifully simple but ancient organisms include bacteria, viruses, fungi, and eukaryotic microbes (cells that have a nucleus) such as yeast. Because the baby's physiological systems are actively maturing during the first few months to years of life, interactions with these founding microbes exert a lasting impact on organ and tissue development.

Skin-to-skin contact between the mother and her baby and breast-feeding help to complete the microbiome seeding process. Both the skin and breast-feeding transfer specific microbes, many of which differ from those transferred during vaginal delivery. In premature babies, skin-to-skin contact is often referred to as kangaroo care, where the baby is carried against the mother's skin. This not only helps with skin microbiome seeding but also seems to help premature infants catch up in their maturation. Changes in the infant microbiome will occur as the baby grows and matures and the baby's sources of food become more diverse.

Breast milk is the ideal food for the baby with few exceptions, one of which is if the milk has been contaminated with unusually high levels of toxic chemicals that could harm the baby. In addition to providing specific immunological factors that help to protect the baby from infections, breast milk is unique in that it contains certain sugars (oligosaccharides) that our mammalian cells cannot digest but that are needed by our microbes. It is specially designed to feed those newly seeded microbes in the baby's gut and help their maturation over the early stages of an infant's life. An indicator of just how important our microbiome is to us is the fact that human breast milk contains foods designed *exclusively* for the microbes. Additionally, breast milk appears to be a source of extra microbes that are transferred via breast-feeding so that it functions as a type of probiotic food.

Breast milk contains several hundred species of bacteria, and these microbes, plus the microbial food (prebiotics) found in breast milk, help to guide maturation of the infant's gut. In fact, breast milk is probably the first probiotic food the baby will consume. The exact composition of the breast milk microbiome differs based on several factors, including whether the mother delivered vaginally or by cesarean section. Not surprisingly, lactobacilli are prominent in breast milk, along with other lactic-acid-eating bacteria. But these are only the tip of the iceberg. Other bacteria, such as *Bifidobacterium* species and *Staphylococcus aureus,* are found as well. Antibiotic treatment during pregnancy or lactation can affect the concentration of bacteria in breast milk. Additionally, the milk of mothers who delivered vaginally had an increased diversity of bacteria, with fewer *Staphylococcus* species bacteria than the milk from mothers who had an elective C-section. Just like other probiotics in food or supplements, the microbes within human milk can alter the baby's metabolism and may even take up longer-term residence in the baby's gut.

In turn, when the gut microbes of the baby are fed their preferred food, they will produce breakdown products (i.e., metabolites) from the breast milk that the baby needs to grow and mature. Again, breast milk contains unique food designed not for the baby's mammalian cells but to be used by the baby's microbes to produce vitamins and other metabolites that a baby needs. Obviously, formulas and other breast milk substitutes that do not adequately feed the baby's newly founded microbiome can alter the course of microbiome development and also can result in developmental problems for the baby's physiological systems. This is something that developers of formulas back in the twentieth century simply did not understand. They were operating under the old biology.

Other body sites of the baby exposed to the environment, such as the airways and the urogenital tract, are also populated with microbes shortly after birth. In general, far more is known about the microbes of the gastrointestinal tract than those inhabiting the other body sites.

This is simply a reflection of the amount of microbiome-related research that has focused on the gut compared with the skin, airways, and uro-genital sites.

As the baby grows and matures, the microbiome grows and matures as well. It is a true partnership, with the microbes of each body site fine-tuned to coexist at that particular site and in communication with those particular cells in the baby's body. Each life stage of the growing child exhibits changes in the physiological systems as well as in the mix of microbes. What happens at these early stages with the microbiome is absolutely critical for later-life health. That is because the baby is very sensitive to being developmentally programmed for gene activity from conception through the first couple of years of life. Those developmental windows, which I have termed critical windows of vulnerability in prior publications, are when attention to the care and feeding of the microbiome can yield the biggest dividends. It turns out that each physiological system (e.g., immune, respiratory, neurological) has its own specific developmental windows of vulnerability that are very sensitive to environmental influences, including those affected by the microbiome. This means that getting a well-balanced microbiome in place early has added health advantages.

Where Did We Come From?

The microbial world is far more than meets the eye. Soil and some plants harbor bacteria that have the capacity to "fix" nitrogen. That means they can take nitrogen gas from the atmosphere and turn it into a form (e.g., ammonia) that plants like peas, soybeans, and alfalfa can use and that eventually enriches the soil. In return the mutualistic nitrogen-fixing bacteria living among the root hairs of some plants get energy sources from the plants. They also cycle the building blocks of protein (amino acids) with the plant, each helping the other out.

The earth itself appears to be encased in a microbial bubble. Recent studies suggest that the range of environmental microbes extends into earth's upper atmosphere under remarkably harsh conditions. In fact, it is thought that they are likely to affect, if not control, climate. Analysis of recent hurricanes showed that the bacterial communities in the hurricane cells were different compared to the regular bacterial composition of the upper atmosphere. Patterns can reveal similarities otherwise overlooked. Hurricanes are a perturbation of the atmosphere, so the pattern of atmospheric microbes is altered from the norm; maybe what happens in humans with the perturbation of the microbiome results in a hurricane in the body in the form of a noncommunicable disease.

One of the current questions is whether microbes can survive in space. A newly discovered, extremely tolerant bacterium has shown up twice in different space agency facilities where highly sterilized materials were being prepared for launch. One case was at the Kennedy Space Center in Florida and a second at the European Space Agency facility in French Guiana. In fact, part of the name given to this new family of bacteria translates from the Latin into "clean." Some evidence suggests that certain bacteria have the capacity to survive the harsh conditions of space. Experiments were conducted on the International Space Station, and spores of a particular bacterium (*Bacillus pumilus*) that had previously been isolated on prior spacecrafts were able to survive real space exposure. The bacterial cells subsequently produced by viable spores had an increased resistance to the most damaging type of ultraviolet radiation.

Whether we superorganisms originated here on earth or elsewhere, it seems clear that our earliest ancestors grew up with microbes as an integral part of their lives. A novel team of diligent researchers from the anthropology, computer science, natural resources, and biochemistry departments of several US universities compared the microbes present in feces samples found in archeological digs of extinct early humanoid communities. They found that not only were the microbial analyses possible, but the results showed these samples matched present-day human microbiomes rather well. However, the similarities were greatest for ancient human predecessors and present-day humans residing in agricultural communities. Urban living appears to have shifted our microbiome significantly from what has been found so far among our most ancient predecessors.

Not surprisingly, as human behavior and food supplies changed in our early existence, so, too, did our microbiome. Scientists in Australia have analyzed the DNA of the oral microbiome from ancient teeth and compared bacterial species across different eras of human civilization. They found the transition from a hunter-gatherer society to one based on agriculture was directly associated with a shift in the types of bacteria found in the mouth. Our microbiome matches our fundamental lifestyle and has for a very long time.

The idea that we owe our continued existence to microbes and will not function well or be healthy without our microbial partners is not totally heretical. The groundwork was firmly laid via the work of famed biologist, National Academy of Sciences inductee, and then Boston University professor Lynn Margulis. Margulis was a visionary in her own right, and she married Carl Sagan, physicist and biologist, Cornell professor, and the soon-to-be host of TV's *Cosmos* series and the most popular scientist of his generation. Can you imagine the dinner table conversations? Margulis and Sagan made quite a scientific power couple, though, in fact, they went their separate ways just before real fame struck.

In 1967 Margulis first suggested the idea that ancient bacteria were so critical for our cells' function that our own cells had captured and incorporated these bacteria into their cellular structure. In the process known as endosymbiosis, different domains of life got intermixed. As mammals with nucleated cells, we are part of the Eukaryota domain of life. Our cells literally ate organisms from the Bacteria domain of life (not surprisingly made up of bacteria) and then kept them inside as part of new hybrid cells. These new cells kept the bacteria, including the bacterial genes, virtually intact. These bacterial remnants are the mitochondria, which sit outside each mammalian cell's nucleus, where our chromosomes live. All cells with a nucleus have mitochondria.

Even plants have an organelle called the chloroplast that is thought to originally have been a type of bacteria. Both mitochondria and chloroplasts generate energy for the cells in ways that are entirely different from the methods the cells use to generate energy. As a result, the new hybrid organism gained both power and adaptability.

The essence of the merged-species idea is summarized in a book written by Lynn Margulis and her son Dorion Sagan titled *Acquiring Genomes*. Margulis believed that the progress of species evolution occurred more significantly by interspecies deals than otherwise. Obviously, this did not set well with strict Darwinian admirers who were looking for more tedious mutation-based development. The question is, why wait so long and hope for mutations when you can beg, borrow, or steal a whole useful genome or at least some advantageous bacterial genes? In fact, there is good evidence that gene exchange happens quite often. Molecular evidence suggests that many of the chromosomal genes in eukaryotes probably originated in archaea and bacteria. In other words, we are chimeras. It would seem that some of the functions found in the human genome originated with our bacterial ancestors.

When you add horizontal transfer of bacterial genes to mammals, including humans, along with the billions of microbes that call the human body home, we become a rather impressive superorganism. We are a holobiont, like a coral reef with its wide diversity of organisms working together to create a whole that is greater than the sum of the parts.

Indeed, maybe we superorganisms were once more like a sort of coral reef warmed by geothermal energy beneath a frozen ocean on a moon orbiting Jupiter than like individual Olympian demigods.

What Can Kill Us?

My roles as a research toxicologist focused on the immune system, director of Cornell's university-wide toxicology program, and senior fellow in Cornell's Center for the Environment required considerable thought about safety evaluation. That is the protection of human health—including its history, present-day status, and future evolution—as well as the broader ecosystem. The fundamental tenet of toxicology and environmental safety in general was voiced back in the 1500s by the German physician, alchemist, and polymath Paracelsus. It is what drove the entire field of toxicology. The mantra is "the dose makes the poison." The real-life effect of this mantra is that what is safe and even useful at one dose might make you sick or kill you at a higher dose. This remains a driving force in modern-day toxicology and is applied through various government-driven safety regulations around the globe. It holds for all of toxicology with only a few exceptions. For example, at the moment scientists and regulators are wondering whether there is truly a safe level for human exposure to some heavy metals such as lead. A safe level of lead exposure has yet to be found as our capacity to measure the adverse effects of lead exposure has increased significantly over the past decade.

What we call safe is only as good as our methods used to evaluate safety. While the science and practice of toxicology has saved countless lives and evolved significantly from the earliest days of food tasters, it is not without historic shortcomings. In fact, the history of toxicology is full of unpleasant surprises and has led to the conclusion that what we don't know can kill us.

Lead in ancient pewter ware and glass, and the human exposure that resulted, is thought to have helped hasten the decline of the Roman Empire. According to the eighteenth-century Scottish physician and chemist William Cullen, in the Middle Ages arsenic was a go-to poison for politically based assassinations. The Borgia family relied on it heavily, and it is thought that Napoleon Bonaparte died of arsenic poisoning. Mercury used in industrial advances led to many unintended consequences. The term "mad as a hatter" is derived from the heavy mercury exposure within the hat industry (millinery) from exposure to vapors associated with the felting process. But other craftsmen were equally involved with unsuspected risks. The advent of silver-plating technology in Britain (e.g., London, Birmingham, Edinburgh) during the early nineteenth century landed many a goldsmith in either an insane asylum or an early grave.

The twentieth-century play *Arsenic and Old Lace* depicted the heavy metal arsenic as a source of homicide, with the story later adapted to a movie starring Cary Grant. More exotic toxins, including those from trees, were featured even in nineteenth-century literature and romantic operas. For example, the tropical manchineel tree with its many toxic chemicals is a major plot element in Meyerbeer's final operatic work, *L'Africaine.* There it serves as a marathon opera-ending method for the lead soprano's suicide. A second toxic tree of literary fame is the Asian upas found in places like Java. Its chemicals can produce heart attacks. It turns out the tree does produce a highly toxic substance, but it usually has to be concentrated before creating the type of widespread killing that captured the literary imagination of Erasmus Darwin, the grandfather of Charles Darwin; the Russian poet Alexander Pushkin; and others.

During my time as toxicology director, I sometimes authored blurbs for the *New York Times* science section Q&A regarding public health toxicology issues. The questions ranged from "Why can you eat blue cheese and not die?" to the toxicity of some fruit pits (e.g., apricots). It turns out the latter make a chemical called amygdalin that, when mixed with stomach acid, produces the poison cyanide. Little did I know that that article would eventually lead to the identification of an imported health food product loaded with amygdalin that had been jeopardizing the health of Manhattan consumers.

Natural toxic chemicals exist as well. Poison dart frogs make a poison used by indigenous populations on their arrow tips. Moldy grains can be contaminated with aflatoxin, resulting in disease and death for those who consume the contaminated food. Given all these toxins in our environment, how come we aren't all already dead?

The Superhero Outfit

Recently, the microbiome has taken on new importance as a type of protective wardrobe that is able to connect us seamlessly with our external environment. You can think of it a little like a Batman or Spider-Man suit. It helps to make you who you are.

It prescreens or filters everything we see in the environment outside ourselves (foods, drugs, chemicals, other microbes), and it is our gatekeeper, determining what gets through to our mammalian cells, tissues, and organs. You can also think of it as our universal translator for a world we otherwise would view as highly threatening. Professor Ellen Silbergeld of the Johns Hopkins School of Public Health and I jointly published a paper describing the gatekeeping function of the microbiome. Other researchers such as Peter Turnbaugh and colleagues have described the importance of the microbiome in interactions with substances in the world external to us (called xenobiotics). The microbiome links our external and internal environments with communication occurring in both directions. If the microbiome is absent, deficient, or defective, our living, breathing, dynamic connection to the world is in trouble. Our very existence then becomes an us-versus-the-environment war with an underdeveloped, untrained immune system as the sole arbiter. The microbiome knows both our insides and our outsides. When it fails, we are left with a system unable to recognize what is us and what is external to us. The consequences are enormous, and we can see them all around us.

In many ways your microbiome should fit like a glove. It can be and should be a perfect match for your mammalian self, such that the two components work hand in hand. As I was preparing this chapter, I came across an analogy from the sporting news.

In the world of competitive athletics, one's garb can help make the athlete. This is particularly true when speed, agility, and/or endurance are involved. Use of skintight suits can provide an aerodynamic advantage while supporting the individual athlete's maximum physical performance. It creates a competitive edge. Space-age technology goes into these competitive uniforms. At the 2008 Summer Olympics in Beijing, the US men's swimming team's special Speedo suits were the rage, and the Spyder-designed suits worn by gold medalist Lindsey Vonn and other US skiers were thought to be an advantage at the 2010 Vancouver Winter Olympics.

But technology in the absence of individual, personalized suitability is not always the answer. Take, for example, the highly favored US speed skating team, which utilized newly designed, specially crafted, high-tech suits requiring that measurements be taken well in advance of the competition. The new suits, called Mach 39, arrived just before the start of the 2014 Winter Olympic in Sochi, meaning that the athletes had not been able to wear them in competition. In contrast, the Dutch team brought competition-proven suits along with their tailor, who would make daily individualized adjustments to the suits and/or equipment. In the end the US team grossly underperformed, causing them to change suits mid-Olympics, while the Dutch team stunned the world with their medal dominance in that sport. If Olympic athletic wear is neither happenstance nor off-the-rack but specifically tailored to fit the individual competitor and ultimately to enhance his or her performance, your microbiome appears to be uniquely matched to your specific mammalian genome. To your body, it should feel like an old friend. From a biological perspective, this makes sense since they have to work together just as life on a coral reef has to in order to thrive. With natural, vaginal childbirth, you would grow from a fertilized egg containing a mixed selection of your parents' mammalian chromosomes as well as the microbes acquired from your mother that lived with your mother's mammalian genes. Human studies have shown how your microbiome complements your host mammalian genes.

One way to examine genetic versus environmental effects in humans is through studies of identical twins, which are babies who come from the same fertilized egg and are genetically identical. Fraternal twins come from two different eggs fertilized by two different sperm, though they develop side by side in their mother's womb. Twins can also share a placenta or each have their own. For scientific studies, identical twins are golden because their mammalian genetics are a known factor.

But if twins are good, triplets may be even better. A study was conducted in Cork, Ireland, looking at the gut microbiome of three sets of triplets. The babies were followed from birth to one year of age. In each set of triplets, two babies were from the same sperm and egg (developing as identical twins and carrying identical mammalian chromosomes) while a third was from a different sperm and different egg (and different in some mammalian genes). This is called the fraternal triplet. All babies were born by elective cesarean delivery, meaning that microbial seeding was not through the mother's vagina. They were fed a mixture of breast milk and formula.

A major focus was on one set of healthy triplets where none of the babies received antibiotics. At one month of age, the microbiomes of the two babies from the same sperm and egg were very similar, while the third (fraternal) baby, who had developed from a different sperm and egg but been carried in the same mother, differed from the two siblings in the fecal profile of gut microbes. By one year of age, these differences had largely disappeared between the three healthy triplets. This finding suggests that our own mammalian genes can have some effect on the microbes that match up to complete us. It is a mini marriage of the two sets of mammalian and microbial genes, at least for the first few days of life.

The triplets where some babies got antibiotics produced a different outcome. In these two sets of triplets, the disruption of the babies' microbiome by the antibiotics had a much greater effect on the babies' mix of gut microbes than did the mammalian genetics (sperm and egg differences). The particular egg or sperm each baby developed from became largely irrelevant in the face of the antibiotic treatment.

For understanding how our mammalian and microbial components fit together in a superorganism, it can be useful to ask questions like, who drives our bus? That question as to who controls our complex body is, at least for me, an unanticipated part of this new biology. John Cryan of Cork and his colleagues recently published a paper suggesting the possibility that our microbiome may act more like the puppet master, a Geppetto, to our puppet, Pinocchio. They describe how microbes can dramatically affect brain function and behavior. Exactly who is in charge, our microbiome or our mammalian self, remains an open question. However, the work of Cryan and others is showing us that (1) our gut microbes can produce staggering, mind-altering effects as potent as any drug's, and (2) these are likely to be useful for future therapeutic purposes.

Mechanisms that demonstrate a psychological consequence to the nature of the microbes with which we share our lives raise an array of unsettling existential questions. Perhaps foremost among them is: How many suicides have microbes caused?

As we consider the rest of the new biology that has emerged and how it applies to our health, one thing should become very clear. The microbiome plays a pivotal role in our possibility for a healthful life. Whether you think your microbes are driving your bus or are simply occupying a majority of seats on your bus, they are part of your personal life's journey. You will soon have the capacity to exert some level of control over your own personal microbiome. What might you do with that?

4

THE INCOMPLETE GENERATION

Birth defects—not a cheerful topic. Think of children with missing or distorted limbs or skin with dark splotches. Shapes, forms, and colors are the way we most envision birth defects. It is a topic especially frightening for parents. The worst consequences often emerge as children mature. Lifelong disability and premature death are often among the outcomes. And many birth defects are presently untreatable.

Birth defects can be genetic in origin, such as cystic fibrosis, an inherited disease that affects mucus thickness and sweat glands, and Down syndrome, a genetic disease caused by having an extra copy of human mammalian chromosome 21. Alternatively, environmental factors, often termed teratogens, can also produce birth defects. One such example is thalidomide, a drug given to pregnant women during the 1950s and '60s to prevent nausea. It produced a host of abnormalities in the offspring, including limb reductions and congenital heart defects. Thalidomide is one of the most tragic examples where a presumably safe medical procedure, in this case administration of a drug during pregnancy, resulted in subsequent birth defects. A second environmental example is fetal alcohol syndrome, sometimes called fetal alcohol spectrum disorders. With this condition children experience a range of neurological and sensory challenges.

It is both accurate and, in the end, highly useful to view microbiome-based, early-life problems with the human superorganism as a new type of birth defect. Babies are being born without a necessary part of their body, never mind how small it happens to be. Let us call such problems what they are: birth defects. Such a diagnosis, complete with its own diagnostic code, might help to open up research funding and encourage clinicians to take the problem more seriously.

An entire network of health-related professionals, including researchers, physicians, and health industry scientists, already are actively seeking prevention and cures for things like asthma, autism, diabetes, and obesity. These are medically sanctioned diagnoses. Microbiome deficiency is no less a problem. When breathing challenges were formally recognized as asthma, resources were marshaled to protect pregnant women and children.

I'm sure not everyone would be comfortable with a formal diagnosis, partly because of fiscal issues. However, this effect can help people lead healthier lives, which is my goal. Medical recognition of microbiome defects would mean that physicians would need to address microbiome imbalances. It would fast-track the formal use of probiotics and other microbe-adjusting strategies within the medical communities.

It is not a hard case to make. Following are definitions of birth defects by some well-respected institutions:

1. According to the US CDC, major birth defects include "structural changes in one or more parts of the body . . . present at birth . . . [that] have a serious, adverse effect on health, development, or functional ability of the baby."

2. The Teratology Society, the oldest society worldwide devoted to studying birth defects, has a broader definition that includes structural, functional, and physiological changes.

3. The US National Institutes of Health, through the Eunice Kennedy Shriver National Institute of Child Health and Human

Development, breaks birth defects into structural, functional, and developmental birth defects. These include nervous system, sensory, metabolic, and degenerative disorders.

4. The March of Dimes, a charitable organization whose mission is to fund research, support families, and help women, describes birth defects as "health conditions that are present at birth ... change the shape or function of one or more parts of the body ... [and] cause problems in overall health, how the body develops or how the body works."

5. For the average person, Merriam-Webster defines a birth defect as "a physical or biochemical defect that is present at birth and may be inherited or environmentally induced."

There are of course many disagreements about what is included and what isn't by these bodies. Since premature babies have a higher incidence of autism spectrum disorders (ASD), the March of Dimes includes ASD on their list of birth defects, while the CDC does not.

All these lists of birth defects are dynamic, not static, since new birth defects are recognized over time and with advances in research and diagnostic technology improvements. One newly defined birth defect is called capillary malformation–arteriovenous malformation syndrome. A mutation in one specific gene produces a condition in the vasculature of the blood-flow system, resulting in several health problems.

Finally, the known causes of birth defects change over time as there are more scientific discoveries. Each year at the Teratology Society Annual Meeting, there is a specific update presented on human teratogens (foods, chemicals, and drugs that can produce human birth defects). This includes the identification of previously unknown and unrecognized human teratogens.

Microbiome-based birth defects could happen by transmitting a microbiome to the baby that carries the microbial genes locking in obesity, diabetes, heart disease, neurological conditions, or one of the many allergic or autoimmune diseases. It is really no different if the defective genes are transferred among the microbial genes or the genes on the mammalian chromosomes. Defects in the microbiome can occur through diet, drugs, oversanitation, stress, and exposure to chemicals that unintentionally kill microbes and damage the microbiome (many not yet known to be toxic to our microbes).

The fact that environmental conditions, both prenatally and surrounding birth, can have a serious impact on the offspring and the risk of birth defects raises some intriguing questions. Would a physician do nothing to balance a pregnant woman's microbiome better if the woman had metabolic syndrome and would donate an incomplete microbiome to her baby? How would you characterize a medical procedure that depletes the microbiome that a pregnant mom will give to her new baby? That is what happens with antibiotic administration during pregnancy. How would you characterize cesarean delivery if it produces a baby who is missing some important part? If medical interventions or the lack thereof result in microbiome deficiencies at birth, then they are in effect producing a birth defect in the baby.

One of the outcomes of the completed self hypothesis as a biological design for health and well-being is the implication of self-incompleteness. Being incomplete as a human superorganism, either at birth or during childhood, can affect everything about our later life. In a 2014 article published in a birth defects research journal, I proposed the idea that self-incompleteness (the lack of being a fully formed superorganism) is equivalent to a type of birth defect, and because this paper was awarded Best Paper of the Year by the Teratology Society, I was given a platform to present this concept in 2015 before the annual meeting of scientists who are most connected to and responsible for birth defects research.

While incompletness of the human-microbiome superorganism is not like traditional birth defects, mainly because it is invisible to the naked eye, it involves every system in your body. If your physiological system does not receive maturation signals from a healthy microbiome, your body will be missing much-needed nutrients, systems will fail to mature properly, and the immune system will be impaired. This is the new biology as it plays out in the human body. That is how it plays out in a baby with an incomplete microbiome.

Derrick MacFabe, an MD and neuroscience researcher at the University of Western Ontario working on the microbiome and autism spectrum disorders, has demonstrated the power of microbes in crafting our socialization and functional capacities. MacFabe has shown that he can make mice and rats become completely antisocial, completely ignore their littermates, and obsess on a ball simply by altering the concentration of a gut bacterial metabolite, the short-chain fatty acid propionic acid.

Production of specific metabolites like propionic acid; butyrate; vitamins B3 (niacin), B5, B6 (active form), B12, and K; serotonin; dopamine; and countless other microbial by-products is one way the microbiome can influence virtually every physiological system and tissue in the body, including the brain. In an incomplete, depleted, or imbalanced microbiome, the metabolites produced can contribute to physiological problems. This happens in the case of immune development when critical metabolites from the bacterium *Bacteroides fragilis* are missing. The defective early-life microbiome results in a defective immune system, increasing the risk of autoimmune disease.

There is a microbiome-NCD-disability triangle connecting the microbiome to NCDs, NCDs to disabilities, and disabilities to the microbiome. To date, the majority of scientific attention has been limited to any couple of points on this triangle, either the microbiome and NCDs or NCDs and disabilities. But in real life it is important to consider the triangle in its totality.

Bryan Love, a professor at the South Carolina College of Pharmacy and father of two children, became interested in the association of antibiotic use and food allergies. He and an interdisciplinary research team pursued the hypothesis that damage to the microbiome from antibiotic use has brought about immune problems and food allergies. As they worked on their research, the scientists realized that it was impossible to teach even a small university class without encountering students with food allergies. This led them on a quest to define the exact linkage between antibiotic use and risk of childhood food allergies and to seek a solution to the food allergy epidemic.

We've already seen a real-life example of the societal impact of a food allergy in the introduction. And food allergies are just one way that microbiome dysfunction and NCDs show up. What happens when you combine all three legs of the triad—microbiome dysfunction, resulting NCDs, and disability? The result is predicted to lead to major societal changes.

Problems with the microbiome set up a myriad of possible NCDs. These often lead to serious restrictions in function and erode the quality of life in patients. The restrictions can be physically obvious, as with certain autoimmune and neurological conditions, but often these disabilities remain largely invisible. Also, the ramifications may not be fully apparent until the baby ages and the neurological, immune, gastrointestinal, respiratory, endocrine, reproductive, and hepatic systems are fully matured and attempting to function in an adult. By that time, the deficits and functional problems usually become apparent and manifest as NCDs. Nevertheless, it is now possible to measure microbiome status using what are called biomarkers. These can be a measure of the microbes themselves, or they can measure specific microbial functions (such as the production of certain vitamins and other microbial chemicals).

Scientists from various biology-oriented disciplines have described the microbiome as a newly recognized organ. You will see it labeled in publications as the "missing organ," although it was never actually missing. We simply did not know it was there. Yet it functions much like one or more organs. Researchers working on hormones see it as another endocrine organ (i.e., like the thyroid gland) managing hormones. Nutritionists, dieticians, and biochemists often see it as a second liver because of its remarkable digestive and metabolic capacities. To immunologists, it is an organ designed to train the immune system, and neurobiologists and psychologists view it as an organ that controls human cognition and behavior. Whether in the role of second endocrine organ, second liver, immune trainer, or neurological control organ, the microbiome, if inadequate following birth, represents a huge problem for us as a superorganism.

The Growing Disabled Population

A recent Kaiser Family Foundation report came at the birth defect/incomplete generation argument from a different direction. This report stressed that NCDs not only cause death but also are a leading cause of disability. This is the same birth-defect-related disability I have been discussing—only from this medical-foundation viewpoint, the disability begins with the NCD, the disease, rather than with a dysfunctional microbiome that leads to the NCD. The same report emphasized that early interventions were likely to be less costly and more successful than waiting until individuals have already developed one or more NCDs during the aging process. World Health Organization agencies have also formally linked NCDs and disabilities under their action plans. Plus, the United Nations General Assembly has linked NCDs and disabilities through its programming and partner initiatives. Microbiome dysfunctions, NCDs, and disabilities are virtually inseparable. This is the nature of our current epidemic and the basis for the rapidly growing disabled population.

Atypical Humans

Two sociology researchers explored the issue of disabilities and perceived boundaries on a whole human, or what they described as different ranges of "humanness," at the University of Maine's Center for Community Inclusion and Disability Studies. Among the intriguing questions Elizabeth DePoy and Stephen Gilson asked were two of particular interest given recent microbiome research breakthroughs: What does it mean to be human? And what are the boundaries of what people might accept as human in an atypical human body?

I suspect that DePoy and Gilson were thinking more of disabilities where the individual is readily seen as atypical. There is some body cue. Or, alternatively, an individual might be using some technology to aid function and that technology or equipment can be seen. But the reality is that disabilities come in all shapes and sizes and categories. Many are not easily recognized by the naked eye and do not currently have special equipment aids. In fact, a microbially incomplete baby or an adult with a dysfunctional microbiome would not look atypical at all. Only if and when certain NCDs developed would it be likely that he or she could be physically recognized as carrying a disability. How do we know this? We already know what a microbially free child looks like. When I was in Dallas just in the early stages of my biomedical education, a special situation arose where a new baby was about to be born that the doctors knew would die if exposed to microbes. So he was purposefully kept microbially free in the attempt to find some way to save his life.

In 1971, there was the case of David Vetter, who was born with a genetically deficient immune system. Exposure to microbes would not help David's immune system because it wasn't there to be trained. Instead, any pathogens among the microbes would kill David. The family had already lost a son to this condition, and the doctors were ready when David was born. He was delivered by cesarean section and placed directly into a sterile protective bubble environment. David was raised for twelve years in this plastic bubble to protect him from all microbes. David had no immune system and no microbiome to co-mature with him and to enable him to function biologically in the environment of the world. He tragically died in 1984 after an attempted transplant led to complications from an undetected virus in the donor cells.

In humans with missing or defective microbiomes there is no perceivable physical difference in outward appearance. Yet such individuals are likely to be metabolically defective and remarkably vulnerable for a range of later-life diseases. This is the dilemma. You don't see your microbiome, so you can't tell visually when it is severely out of balance. But the defects are still there whether you see them or not. In David Vetter's case, it was only obvious because his immune defect required that he be removed from the world's normal environment and segregated into a completely artificial environment just to stay alive. The presence or absence of a microbiome was not a physically altering body feature.

Despite progress in disability inclusion in many areas of society, the ongoing NCD epidemic and the NCD-disability axis raise serious challenges:

1. NCD-based disabilities inherently restrict access to the full environment normally enjoyed by others.
2. The sheer numbers of those with NCD-driven disabilities mean there are fewer humans who can move across different environments with safety.

The danger is that increasing numbers of us may have severely restricted environments in which we can safely function.

The issue facing society when many people struggle to interact safely with different environments was driven home to my wife and me personally. Before becoming a science editor, my wife was the learning disabilities specialist at the State University of New York at Binghamton (SUNY Binghamton). Her job was to design and implement a transition program for students with learning disabilities, attention deficit hyperactivity disorder (ADHD), ASDs, and traumatic brain injury. When she began as a graduate intern in 1994, there were only six students and the job was both rewarding and manageable. However, numbers of students increased rapidly, first with a jump in students with ADHD, followed by the twenty-first-century explosion of students with Asperger's syndrome.

Unlike students in other categories, students with Asperger's require much more intensive, personal management because they have trouble navigating different environments appropriately. Young adults with autism spectrum disorders such as Asperger's syndrome have a difficult life ahead, usually requiring many services from others. A big question is: Who will be there to provide all of the support services as the population of the needy increases? According to the Asperger/Autism Network, Asperger adults often require assistance for routine living tasks such as paying bills, keeping the house clean, or general organization. They may share living space with assisting housemates if not living directly with parents.

Universities are trying to catch up to the epidemic. However, more trained staff for the increasing numbers of students is not a long-term solution to the challenge. The growth in the disabled population of young adults became so significant that in Canada universities have begun to specialize in the care of specific populations such as those with Asperger's syndrome. Consider that this is just one category of disability. This is happening over and over again across all the physiological systems (immune, neurological, gastrointestinal, cardiovascular, metabolic, respiratory, urogenital) linked with human self-incompletion. In the United States, the population of students with disabilities doubled compared with the growth of the general student population between 1980 and 2005, and during the 1999–2000 school year the US spent $77.3 billion on education and services for the disabled student population.

But it is not just about the expense. My wife's experience is a microcosm of the broader concern about the growth of the disabled population relative to the availability of caretakers should the NCD-disability epidemic continue. As early as 2001 a US Department of Health and Human Services report sent up red flags about disabled youth and the workforce availability of long-term caretakers. We need to find a solution for human incompleteness and the NCD epidemic before we run out of caretakers.

Social Fracturing

As we are stricken with more and more NCDs that make daily life more challenging, we have a tendency to hunker down, to cope with limitations, and to protect ourselves. Individuals may even have to withdraw from what used to be routine social gatherings and interactions with friends, family, and business colleagues. Remember Lord Simon, a member of Britain's Parliament? Despite being a politician, he has to avoid any place where people may congregate because of his extreme allergy. It may seem merely an inconvenience, but what opportunities are being lost if a dedicated member of the government can't freely associate with his or her peers?

Holiday dinner celebrations, wedding receptions, community dinners, summer picnics, conference meals, and even single-family meals are increasingly affected. The host must anticipate the food allergies, intolerances, and medically related dietary restrictions such as those that occur with diabetes. Additionally, cooking utensils and surfaces may need to be duplicated and kept separate. All ingredient labels have to be kept as well for potential scrutiny. For the host, this is becoming an increasingly complex challenge. We are all worried about food in a way that previous generations didn't need to be.

A college student leaves home and heads off to the university and a meal plan full of cafeteria food—oh no. Now she is missing classes due to allergic reactions. One student may need to eat things that would likely kill another student. Another student's health-supportive classroom or dorm room is a hospital visit waiting to happen for another student.

Families may adapt to the mix of food-related and other environmentally related health issues among their members, but when allergic, intolerant, and dietarily restricted children, as well as others spanning the spectrum of NCDs, head to K–12 grades and then on to university, the challenges continue. In 2009, the US Department of Justice received a complaint against Lesley University in Cambridge, Massachusetts, for an incident related to food allergies and celiac disease. The complaint argued that the broader category of food allergies falls under Title III of the Americans with Disabilities Act. The resulting settlement agreed that food allergies are a disability and should be handled as such by the appropriate university student services offices. This new ruling is being interpreted by other universities to extend to them as well, and they are changing campus dining practices and incorporating the new set of NCD-based disabilities into accommodation plans. Obviously, this is a new cost in human capital, our capacity to congregate around a meal, and a type of freedom humans used to have. It is directly tied to problems with the microbiome, gastrointestinal perturbations, and improper immune maturation.

Various factors have driven both passive and overt physical segregation of people in the course of human history. The most significant of these have been race, religion, lifestyle (e.g., agrarian versus nomadic societies), politics, and wealth. Even centuries ago, genetically related clans and tribes tended to control their own territories and stay separate from other clans or tribes. Despite efforts to mingle groups, you can still see this today in many areas of the world, such as under the caste system in India or at the previously mentioned leper colonies.

However, to date, separation by disabilities has been limited to two main groups: seniors, who often have functional limitations accommodated at residential living facilities and communities, and children and young adults with autism spectrum disorders, served by special local schools and residential living complexes. These were established to efficiently provide needed, expert caregiver support to these groups. However, several authors have pointed out that efforts to provide special services and/or special education can unintentionally increase segregation. Despite laws that promote the inclusion of children with disabilities, it is still a challenge. If people are increasingly unable to join one another in the same local environment (e.g., a school or classroom) because of hypersensitivities, segregation could well increase.

Consider what would happen if allergies, autoimmune diseases, metabolic, neurological, and behavioral disorders didn't occur or if functionality could be significantly restored among those carrying these burdens. It could and would reduce what will otherwise become an ever-increasing divide among humans. We have been social animals for thousands of years, and steps that we can take to become more resilient, less codependent, and more capable of broader human interactions would benefit not just us but also the whole world.

5

GENE SWAPS AND SWITCHES

The previous chapters of this part of the book emphasize how the vast majority of your cells and genes are microbial, not mammalian, and why that is important for understanding yourself and your well-being. In particular, the fact that more than 99 percent of the genes in your body are from your microbiome and not your chromosomes is one of the more eye-popping findings in this field of inquiry. What does that mean? A product might be 99 percent lactose-free. That is usually good enough for most lactose-sensitive people to avoid problems when consuming the product in moderation. But are genes different than lactose? Do you really think that the 1 percent of mammalian genes exerts more control than the microbial 99 percent? As we will see, the answer is probably not, considering not just the raw numbers but also a variety of interactions hidden from our normal view. The two genomes, mammalian and microbial, work together. Sometimes it can even be challenging to be sure exactly what is a result of microbial activity and what is mammalian. All of these cells and genes have very ancient origins that are somewhat murky, but definitely interdependent and intertwined.

There are two levels of genetic control involving the microbes and the genes on our chromosomes. Let's call them gene swaps and gene switches. A gene swap is essentially about a gene's location and who possesses it, where it came from, and where it went after the swap. A gene switch has to do with a gene's use, whether a gene is on or off like a table lamp. Gene swaps and switches are a fundamental way microbes exercise their power within you, the superorganism.

Swap

One of the major recent findings in biology is that genes can be swapped. Who would have thought? The very things that we believed made us exactly who we are and distinguish us from others are actually at a type of swap meet. They can be sold or given away like items at a yard sale.

Researchers studying gene swaps try to determine where a gene originated—was it swapped? It is a little like looking at our current genes and asking what spectrum of ancestry gave rise to all of those genes. I like to think of it as looking at the people in the United States today and asking: Where were the prior homes of all of our ancestors? The US is a melting pot of many populations who over centuries migrated from other areas, countries, and continents to this particular geographic location in North America. Of course, it is still happening. A majority of the ancestors of people now in the US originally lived in different parts of the world and, of course, most of their genes were from elsewhere as well. People can move and relocate. And, it turns out, so can genes. The theme of this chapter might well be location, location, location.

What we identify as microbial or mammalian in origin is perhaps the first question. As explained in a prior chapter, a big biological issue of the latter twentieth century concerned the bacteria-like energy powerhouses called mitochondria, which are located in the cytoplasm, a region of our cells that is outside the nucleus. The present consensus is that mitochondria are a remnant of what were once microbes that our ancestors' cells somehow captured because the mitochondria were useful and could diversify our energy sources. Diversifying energy sources can be beneficial, and that topic is in the news today on a larger scale as countries seek to develop renewable energy sources to protect the earth. If the mitochondria were originally microbial, then the genes within the mitochondria were originally microbial genes. But even if these outside-the-nucleus, bacteria-looking cell components were originally bacteria, the nucleus in our cells was surely 100 percent mammalian through and through. Our chromosomes would not be compromised by interspecies sharing or transfers. Or would they?

The main way our chromosomes acquire a gene from a different species is through a process called horizontal gene transfer. This is a form of swap where a gene in one species is snatched or grabbed by a different nearby species (like two species inside a superorganism). Often this exchange seems to advantage one species quickly and the other species slowly. It can be somewhat like when a bank provides a lump sum to pay for a borrower's new house (initial advantage to the borrower), but then the homeowner must pay off the mortgage including interest over decades (longer-term advantage to the bank).

With horizontal transfers, a gene gets moved from one living organism to another during the same generation. Genes as property are exchanged. This horizontal transfer is in contrast to vertical gene transfer, which occurs during reproduction. With vertical transfers, genes are transferred between generations from parent to offspring. In vertical gene transfer for humans, a mother and father transfer chromosomes via the sperm and egg to form the zygote, which grows into the baby. Also, the mother donates her microbiome to the baby at birth as the vertical transfer of microbial genes. Vertical gene transfer has long been recognized and was indeed widely thought to be the only way genes made their way across generations. Horizontal gene transfer is a whole new ball game, at least for science. It requires genes to jump. It may seem like the equivalent of a simple handshake between species, but the actual process is probably a little more mysterious.

In 1950 Cornell-trained geneticist Barbara McClintock showed that genes could be mobile and could indeed jump and change locations, at least along chromosomes within in a cell's nucleus. It took decades before her revolutionary, Nobel Prize–winning discovery was fully embraced and appreciated. But if genes could do that, could they jump between species?

An early result was reported in the 1950s, showing this was possible among bacteria, including the one that causes the disease diphtheria, *Corynebacterium diphtheriae*. In the case of the diphtheria-causing bacterium, genes that were transferred into the diphtheria bacterium by bacterial viruses called bacteriophages controlled how aggressive (or virulent) the bacterium was in producing the disease. Following this, it was shown that the genes providing resistance to antibiotics could be horizontally transferred or swapped between different species of bacteria with the help of the same viruses.

It turns out that our bodies are perfect locations for horizontal gene transfer. In fact, the microbes within our microbiomes are known to use locations such as the gut as a type of swap meet. We only recently discovered that different bacterial species living inside us in the same body location can occasionally exchange genes. But can this type of transfer happen when the recipient is a higher organism: a plant, an animal, or even a human?

The subject of horizontal gene transfer in higher organisms, including humans, has been debated for more than a decade with speculation as to whether a swap or transfer of genes could occur. Horizontal transfer of genes between two plant species, rice and millet, was demonstrated in 2005. Rice is nice, but what about humans? Could genes that originally came from microbes not only end up in human chromosomes but also be transmitted from parent to child during reproduction? In one of the best studies to date, a team of researchers led by Alastair Crisp at the University of Cambridge focused attention on human genes that share remarkable similarities to those of bacteria, archaea, and fungi.

Tens to hundreds of foreign genes of probable microbial origin have been identified in the human mammalian genome, and many of these seem to code for proteins with unique enzyme activities. Because of these functions, these apparent microbial genes appear to provide our cells with chemical-processing capabilities they would otherwise lack. The discovery of microbial genes in our chromosomes raises several questions: Does horizontal gene transfer and genes jumping between species impact Darwin's view of evolution? Can the "tree of life" depicting species relatedness and the process of evolution continue to exist as a pristine tree, or is it really something different? Maybe it is closer to a pecan tree completely encased in webs of the fall webworm (*Hyphantria cunea*). How much of it is tree and how much is caterpillar web depends upon one's perspective.

This microbe-to-human horizontal gene transfer is a relatively new swap discovery, and not everyone is totally convinced horizontal gene transfer is the only explanation for the findings. But the evidence for gene swaps between microbes and other plants and animals is so strong that to exclude humans from this widespread biological process would seem to be a stretch and require us to make the assumption that humans don't do things biologically the same way as most other animals. Most of the debate now is more about when such transfers occurred and into what combinations of vertebrate species.

A prior swap of genes from microbes to our ancestors' mammalian chromosomes would mean that some of the approximately 1 percent of human mammalian genes are not really mammalian at all. At least some of those genes sitting on our chromosomes today were swapped into us from microbes. So the more we look, the less of us is actually nonmicrobial and free from microbial influence. If these once microbial genes help us to do useful stuff that we could not do before we grabbed them, that stuff we now can do is of microbial origin, even if the capacity resides in our own chromosomes. Gene swapping and the microbially originated genes sitting within our mammalian cells make the boundary between the mammalian part of us and our microbiome very fuzzy.

Switch

It turns out that he who controls the gene switch controls a lot. The idea that your genes determine not only who you are but also your appearance, personality, and health profile had much currency as the Human Genome Project was being completed in the 1990s. You probably heard talk of the crime gene, the gay gene, even the intelligence gene. These features of a person add up to what biologists call a phenotype, which is simply a group of observable traits in an individual. The traits might be something you see, such as eye color, height, and facial structure, or something you don't see outwardly but can measure, such as heart size, level of thyroid activity, metabolism, or biochemistry. Biologists have known that the inheritance of genes and different forms of genes, called alleles, does not always predict phenotype. This had been chalked up to interactions between genes and some environmental effects. That's the old biology. We are now realizing that simply having a gene determines very little about how, when, and to what extent you may ever use that gene. The real control is whether a gene gets switched on and when. In most cases, if it just sits there on a chromosome and is unused, it might as well not be there. The control of gene use is called epigenetics, and this control mechanism is a central component of the new biology.

As mentioned in the introduction, humans have an underwhelming number of mammalian genes that, by themselves, are not capable of sustaining human life. That is why our second genome, via the microbiome, is not simply a luxury but a necessary and fundamental part of our being. However, genes are a little like electricity in the modern world. You can do amazing things with it, it likely powers your house or apartment and maybe even your car, but it is only useful once you can plug into it and control it—i.e., turn it on and off.

Wiring your house or apartment for electricity is only a potential for use. It provides a potential for having light and using electrical appliances. However, you need a circuit box with circuit breakers and light switches as well. You also need access to the electricity via outlets. If electricity simply comes to your house but you lack circuit breakers, light switches, and outlets, you are not going to see any benefit from the initial wiring. You have no access; only a potential exists. Genes are the same way. Whether they are mammalian or microbial, or genes from outer space, it makes no difference if they can't be turned on.

We are very fortunate that, just as switches in a house can be installed by an electrician in keeping with code, our genomes come with both access and switches. The only difference is that the switches are not toggles to be physically flipped. Instead, they are chemical switches. And there are several different kinds of chemical switches. Understanding and better utilizing these chemicals is part of the new biology and the future of human medicine.

Being able to control when a gene is turned on, how much of a gene's product can be made, and when during development the gene is on or off can be the difference between life and death, health and disease. Consider the production of hemoglobin, the oxygen-carrying protein in blood. Without adequate oxygen, your cells and tissues would die. It turns out you have different types of hemoglobin, and they are tailored precisely for different life stages and the oxygen needs of your tissues during those specific life stages. The production of embryonic versus fetal versus adult hemoglobin is under the control of epigenetic gene switches. The switches are flipped at precisely the right developmental stages for everything to work. It turns out that one of the small molecular metabolites of gut bacteria, sodium butyrate, can control these gene switches and affect hemoglobin expression. It and related chemicals are being tested for possible use in treating hemoglobin-related diseases such as sickle-cell anemia and beta-thalassemia. In those diseases, tissues often do not receive enough oxygen. Sodium butyrate can boost the amount of a high-oxygen-carrying form of hemoglobin in the blood. Clearly the microbiome has a biological role in the control of gene switches.

I am not alone in thinking about epigenetic control of gene usage as a type of switch. Recently, Dr. Dietmar Spengler and colleagues at the Max Planck Institute for Psychiatry in Germany described how chemical switches for gene usage are critical for healthy neurological development. They also described what can go wrong with the programming of these switches. They used the analogy that these switches are a part of writing your very own personal book of life.

I like to think about the programming of the switches for our development much as you might program the lights in your house while you go on a week's vacation. In the old days you might have set this up using timers plugged into electrical outlets. Today, they could be computer-driven and connected in "smart houses." If you had only one chance at the programming and it would last the entire week of your vacation, you would need to get the programming right so that lights in different locations in the house and yard would come on when they should and go off when they should for maximum security. If you got the timing wrong, it wouldn't work out well, and lights could turn on during daylight and turn off at night. This is what can happen with gene switches in your body—with far worse consequences.

You can also think about these switches across different periods of life. A good analogy is that of railroad switches that determine the track taken by a train. For example, the longest rail line in the world, at approximately 5,772 miles, is the Trans-Siberian Railway. It connects Moscow through the Ural Mountains with the Russian Far East, the port city of Vladivostok, and the Sea of Japan. Shortly after arriving at Lake Baikal in the east Siberian town of Ulan-Ude there are track switches. The main track follows a route (the Trans-Mongolian Route) that leads through Mongolia (Ulaanbaatar) and into China, eventually reaching Beijing. A later switch point in eastern Siberia is about sixty miles past Chita, where a line switches off and runs directly southeast to China, leading to Beijing but skirting around Mongolia. These switches send trains to different regions.

Complex biological functions are under a certain level of gene switch/epigenetic control. These include such critical human functions as the formation of and maintenance of memories, the effectiveness of your immune response, the levels of specific hormones in your body and your responses to those hormones, and the levels and quality of sperm production.

The point is that these switches matter so much because they can be programmed. The programming begins in early life but can also occur during the lives of our parents and grandparents. In effect, they connect us both to our past and to our probable future. And of course, in some cases, the microbes in our microbiome can tell our mammalian genes whether they should be on or off both in the moment in real time and also later in our life and even in our grandchildren's lives.

These switches, also called epigenetic marks, have their own memory. The memories of these gene switches are just as important as any of the mammalian and mitochondrial genes we have inherited. These epigenetic "memories" can span generations.

Giraffes' Necks

One of the most remarkable rebound stories in the history of biology, if not all of science, has been the changing fates of biologist and naturalist Jean-Baptiste Lamarck. Before Darwin there was Lamarck, whose theory of evolution said that environmental adaptation was driving generational changes and eventually inherited traits. Essentially, giraffes' necks are long because adult giraffes stretch their necks when reaching for leaves high in the trees. These long necks acquired in maturity can be passed along to their progeny. This was at odds, to say the least, with Darwin's approach.

Lamarck was born Jean-Baptiste-Pierre-Antoine de Monet, chevalier de Lamarck, in 1744 and grew up in a large family in northern France. He distinguished himself as a military officer until an injury forced his retirement. After that he began studying medicine and botany and produced a well-received book on the plants of France in 1778. He was appointed as a natural science professor but in an area viewed as comparatively lowly at the time, the study of invertebrates: i.e., insects and worms. It was through his work on the diversity of lower animal life forms that Lamarck began to form his views on adaptation. He believed that environmental influence on organisms could produce long-term effects as the organism underwent changes in the use of its cells, tissues, and organs. And so he reached his conclusion that, when the interactions spanned significant time, these changes could be inherited and observed across generations.

Lamarck had exceptionally broad scholarly pursuits and writings. His interests spanned medical science and botany and even extended into physics. Nevertheless, he died in poverty and obscurity. Only in the past few decades, as scientists have begun to discover the significance and impact of epigenetics, have Lamarck's theories been reexamined. A few previously dismissed ideas have gained new relevance. And those ideas are now at the forefront of how we're considering human health protection and how we'll treat disease in the coming decades.

The idea of inherited changes through environmental adaptation does not seem as absurd as it once did. In the twentieth century, when science focused on inherited genes, including a rediscovery of Mendel's work with peas, Lamarck and his ideas were ridiculed if they were paid any attention at all. He had become a poster boy for wrongheaded biological thinking by the time I was in school.

Yet as we have come to see over the last decade or so, what Lamarck described, environmentally driven adaptation, is precisely how epigenetic regulation of gene expression appears to work. He did not have the tools we have, but his perspective is profoundly valuable today. It is a good lesson in how scientific consensus can blinker us to new ideas and breakthroughs in our understanding.

The Adult Health Program

To stay healthy we need to stick to the epigenetic program. If genes we need to switch on for a certain piece of our development fail to switch on or switch on at the wrong stage of our life, it usually causes disease. The resulting diseases that show up are most often of the noncommunicable type.

This process of establishing the pattern of gene activation while you are a baby is often referred to as developmental programming. It is much like programming a computer to do a virus check once a week in the middle of the night when you don't need to use your computer. Much of my own career has been devoted to questions of how, when, and where programming for your developing immune system occurs. The genes on your chromosomes can be programmed based on early-life environmental exposure, including maternal and childhood diet, exposure to hazardous chemicals and certain drugs, or the presence or absence of key microbes. Each physiological system of your body undergoes this type of developmental programming. For some physiological systems and organs, full adult-level maturation happens earlier in life than in others. For example, the brain and lungs are among the last to reach full maturity as you age.

Developmental programming of later-life disease was first discovered about 1990 by UK researcher David Barker as he studied the developmental basis of heart disease. Barker noticed that if mothers had a limited supply of food, their offspring's growth curve changed, and the children were more likely to develop metabolic problems, including heart disease. His theories on developmental programming of heart disease became known as the Barker hypothesis.

Through additional investigation conducted by scientists such as Philippe Grandjean of the University of Southern Denmark and the Harvard School of Public Health, Cheryl Walker of the Texas A&M Health Science Center, and Jerry Heindel of the US National Institute of Environmental Health Sciences (NIEHS), we now know that additional NCDs also follow the same pattern of early-life developmental programming of gene activation, usage, and risk of later-life disease.

The new biology covering the developmental programming of human health has become so extensive that brand-new scientific societies and research journals committed to this topic have popped up during the past decade.

The ramifications of gene switching have utterly befuddled the great biological debate of the twentieth century: nature versus nurture (genetics versus environment). As suspected for years now, the two can no longer be usefully separated. That paradigm is simply outdated. The environment so controls the programming of which genes you use that what you are from cradle to grave reflects in large part your combined ancestral and early-life nurturing and experiences (e.g., chemical, physical, and psychological stressors). This was eloquently described at the molecular level by David Crews (the University of Texas at Austin) and colleagues. The researchers described how continued focus on nature versus nurture (part of the old biology) is a problem because the "hoary concept of evaluating traits according to nature versus nurture continues to persist despite repeated demonstrations that it retards, rather than advances, our understanding of biological processes." We need to move beyond this to advance our biological understanding of humans as a complex, yet fully integrated, superorganism. The food your ancestors ate (or didn't eat), the air they breathed, and the water they drank all affected their on-off genetic switches.

These gene switches are now playing out in you across your life span. Your gene switches appear to have a memory of what your recent ancestors encountered in terms of stress, food, chemicals, and drugs. Of course this can make it challenging to know whether an environmental effect we see in our lives now is due to a present-generation environmental exposure or something that our parents encountered that still controls our on-off switches.

Evidence for this epigenetic memory, also called transgenerational environmental epigenetics, exists not only in lab animals, through the work of researchers such as Michael Skinner (Washington State University) and Andrea Gore (the University of Texas at Austin) on endocrine-disrupting chemicals in mice, but also in direct human experiences.

Two prime examples of epigenetic memory in humans involve the Dutch famine of 1944–45 (also called the *Hongerwinter*, or hunger winter) and the Great Chinese Famine of 1958–61. Apart from the deaths caused by immediate starvation, there were effects in the descendants of those who survived. Ironically, neither of these famines was caused by weather changes affecting crop production and subsequent food availability. They were a direct result of human action: politics and war. The Dutch famine occurred near the end of World War II because the Nazis blocked shipments into the occupied part of the Netherlands. It continued until the Allied forces did food drops in early 1945 prior to the liberation of the area.

While tens of thousands died of starvation, the generational epigenetic effects that lingered were a scientific surprise. The Dutch Famine Birth Cohort Study has provided an opportunity to evaluate the effects of this war-induced famine. Babies developing in utero during the Dutch famine were found to have what are called epigenetic marks for genes involved in metabolism. There is evidence suggesting that the packaging of the DNA in the babies' chromosomes was altered by the environmental conditions of the famine. This, in turn, affected the expression, or switching on of those genes, and the babies' metabolism later in life. For example, babies exposed in utero to the famine conditions were more susceptible to the development of type 2 diabetes as adults. The risk of developing diabetes was directly related to the severity of the famine the developing babies experienced. Those whose mothers suffered more severe malnutrition had the greatest risk of developing diabetes as adults. There is even evidence that it carried on to the next generation. Remarkably, the children of fathers who were exposed while in utero during the winter of 1944–45 were heavier and more obese than the general population.

Most scholars agree that the Great Chinese Famine of 1958–61, which led to approximately thirty million deaths, was a man-made catastrophe caused by massive changes in agricultural and other policies during Mao Zedong's Great Leap Forward. It is considered to be the largest famine in human history based on the number of people affected. While it produced similar results to the Dutch famine, Mao's redirection of food to cities and the complete lack of food in rural areas allowed for some interesting comparisons of long-term effects among large numbers of people within the same country. A significant number of studies have been conducted on the offspring of famine survivors within the past decade.

Because the Great Chinese Famine is more recent than the Dutch famine, more information is available on the epigenetics and health of the children of the Chinese survivors than on those of their grandchildren and great-grandchildren. However, the developmental programming effects on the health of the Chinese offspring are clear and quite sobering. Among the noncommunicable diseases and conditions at elevated occurrence in the Chinese descendants of the hardest-hit areas of the famine are metabolic syndrome, schizophrenia, and anemia.

These two examples suggest what has been reported in mice and rats—that the effects of nurture on DNA packaging and a gene's on-off switch as programmed in early life can affect later-life risk of noncommunicable diseases. Additionally, at least some of these thrown gene switches may be preserved and transmitted to subsequent generations that were never exposed to the actual environmental conditions.

If epigenetics, the control of your on-off gene switches, is one of the more remarkable biological discoveries of recent decades, there is still the question of how the microbiome fits into the picture. This is where it gets really interesting. In a previous chapter, I discussed the primary role of the microbiome as your gatekeeper. It serves as a type of protective bubble for you that filters all of your environment and determines what actually reaches your human mammalian cells. It does this for what you eat, breathe, and come into contact with, whether food, environmental chemicals, or drugs. If the environmental exposures of your cells control your gene switches and your microbiome filters your environmental exposures, then guess what exerts a massive effect on your gene switches? Your microbiome. In a sense, your microbes—and which specific bacteria, viruses, and fungi you have in your gut, reproductive tract, and airways and on your skin—have a significant effect on the gene-switch-throwing chemicals that your cells and the mammalian genes in your chromosomes see.

A recent discovery on the gene switches puts the microbiome not just in a passive role as environmental filter but also in the active role of master controller for the switches. It turns out that many of the metabolites released by our microbes can flip the switches in many of our mammalian genes. The microbiome is a major player in establishing our developmental programming, in part through control of the gene switches. Having a complete and healthy microbiome in early life is critical for the healthy programming of the genes in our developing physiological systems.

This is a warning call about the long-term effects of microbiome depletion. If microbiome-related epigenetic marks are transmitted across generations, the full range of effects, including those on subsequent generations, could be harder to correct than by simply taking a probiotic pill.

The first part of this book introduced a new way of thinking about biology in general and human biology in particular. This new biology will revolutionize medicine, human health protection, and opportunities for improved self-care. The philosophical implications are beyond me, but they are nothing less than thinking of yourself as a tropical rain forest or a coral reef. That is at least somewhat existentially unnerving.

To your health!

PART TWO

A REVOLUTION IN MEDICINE

6

REDIRECTING PRECISION MEDICINE

Medicine is only as good as the biological underpinnings that support it. Change the biology, and you are forced to change the medicine. You can see this happening as new biological findings gradually work their way into university biomedical curricula, continuing medical education courses used to update physicians, and pharmaceutical, nutrition, public health, and public safety conferences. Of course that can be a slow, torturous process before we as patients see any tangible benefits. Changes are happening in medicine today. But are those changes actually aligned with our latest understanding of fundamental human biology? To the question of "Are we there yet?" my answer is a resounding NO. The most recent medical initiatives remain rooted in a fundamentally flawed concept of human biology. That flaw is the premise that our mammalian genes drive our health most significantly, and much of medicine remains wedded to that premise. Even recent major initiatives are more status quo than meets the eye.

In this new era of the microbiome, there is a massive gap between what we know about human biology and how human health is managed in westernized medicine. That gap needs to be closed soon so medicine can look more like the ecological management of a coral reef than something akin to trying to smother a forest fire with a small blanket.

Recent changes in medicine, while well-intentioned, are largely misdirected. Two new medical initiatives that are changing the playing field are "personalized medicine" and "precision medicine." You may have come across one or both of these terms in some context. Essentially, they are two parts of the same initiative to orient health care toward the individual. A focus on the patient as a unique person rather than just one part of a larger population appears to be the inescapable future trend in medical care, and that is a very good thing. But with these new medical initiatives, the devil is in the details when it comes to how personalized and precision medicine are focused practically. In this case the problem is what is meant by "the individual." You could easily assume it means all of you, the whole individual, a holistic view of you as a unique superorganism. But the reality is it means a form of medicine that remains focused on a very small portion of you: largely the minority, human, mammalian genome.

Personalized medicine as a conceptual force arose around the beginning of the twenty-first century, although the foundations for it go back several decades. It emerged on the heels of the Human Genome Project and focused on the idea that minor mammalian genomic differences (estimated at only 0.9 percent) among humans were vital to our better health. If only we could treat according to these individual chromosomal differences, we would better personalize our medicine. The intentions were good, and the goals were to achieve cost savings while providing better medical solutions. Of course, remember the previously discussed, inconvenient fact that the Human Genome Project resulted in grossly underwhelming results from what had originally been anticipated.

Precision medicine is largely an extension of personalized medicine, and many use the terms interchangeably. Precision medicine was launched as a US initiative in January 2015 with a presidential announcement followed by a prominent joint announcement in the *New England Journal of Medicine* by the head of the National Institutes of Health and the former director of the National Cancer Institute.

Similarly to personal medicine, precision medicine emphasizes the individual variability of your genes, environment, and lifestyle exposures, and the use of that information to improve disease prevention and treatment. The initiative emphasizes using lots of data, or what has been called "data-intensive biology," as a way to see trends in diseases and treatments and how all of the pieces fit together in the individual patient. In effect precision medicine will link your genetic, environmental, and biological information to your electronic health records.

The near-term focus of precision medicine is on a single disease category, cancer, and in particular the identification of human mammalian genes that drive the development of tumors. A second priority is the use of networked technologies and social media for better patient diagnosis, treatment, and care and to respond to an increasing trend of Americans wanting to engage medical researchers as active partners. Not everyone sees precision medicine as a magic elixir. Among the criticisms of this practice is that it forces a very reductionist view of human health. If the keys to better health are not where precision medicine happens to be focused, that's a big problem.

Of course there is one glaring weakness in the personalized medicine/precision medicine strategy, and that is the heavy focus on the human mammalian genome. The problem is right there in the numbers. We are focusing our medical initiatives on only 1 percent of our total superorganism genome. The math is easy. It means focusing medicine on fewer than one-hundredth of all our genes. Something is very wrong with this picture.

If our microbiome has approximately 99 percent of our genes and is readily adjustable, why would we focus on what is less than one-hundredth of our genes just because they happen to be sitting on our mammalian chromosomes? What about the 99 percent of us residing among the microbes of our gut, airways, or skin? Add to that the problem that the gene switches (epigenetics) control much of the activity of the genes on those mammalian chromosomes, and it would seem that we might be gearing up to look in the wrong place for future medical solutions.

To be fair, the precision medicine announcements do include a mention of gut microbes and the potential importance of fecal sampling and individual analysis of the patient's microbiome. But this seems more like an afterthought for future data collection than a primary focus of this new medical initiative. The stunning impact of the human microbiome project has been left largely out of the latest medical initiatives. But that does not mean things cannot change as more and more patients and physicians embrace their true biological nature.

Despite the mammalian-centric focus of precision medicine, many in medicine, including drug companies, the allied health industries, and physicians, are not ignoring the microbiome. Between 2014 and 2015 alone, I presented at a wide spectrum of biomedically related conferences involving pediatric MDs, OB-GYN MDs, autism researchers and clinicians, the pharmaceutical industry, nutrition and food company representatives, probiotic researchers and clinicians, birth defects scientists, and health and safety overseers. It was an interesting personal challenge to prepare lectures for groups with such widely divergent concerns. However, the common denominator for all of these conferences was the microbiome. Even sessions at these conferences not specifically dealing with the microbiome tended to drift to this topic once the Q&A session began. It is the 800-pound gorilla at biomedical conferences, and it is soon to be that gorilla in doctors' offices.

If your health care providers are not talking with you about the microbiome and asking about your intake of probiotics yet, it is likely they will be by the time of your next annual visit. It is not something to be ignored. Among many reports of its kind, a recent *Los Angeles Times* article detailed recent progress on the microbiome that will enable doctors to both fingerprint you based on the microbiome and make designer adjustments to your microbiome. It is becoming apparent that few doctors are going to want to be practicing twentieth-century medicine as we move further into the twenty-first-century era of the microbiome and human superorganism.

Part of any prognostication about when, where, and how medicine will soon become truly holistic and treat the entire human superorganism involves the patient. In a sense medicine is still a service industry. Physicians provide a service and you are their customer. If antibiotic overprescription in doctors' offices can be linked to pharmaceutical representative visits, then patients' expectation of leaving the office with something, such as a prescription for antibiotics, is also a factor. But turn that around and consider patients beginning to expect microbiome-based medicine to be a part of the visits. Think about patients expecting doctors to inquire about or evaluate their 99 percent, their microbiome status.

The patient-doctor nexus generates a powerful social force, ultimately driving the landscape of medicine, and you play a key role in this. Take, for example, recent information from the Arthritis Foundation on the patient-doctor visit. They emphasize that during a standard eighteen-minute visit of a patient to a primary care physician in the United States, doctors have a checklist of things to accomplish. Patient preparation and prioritization is critical and impacts eventual satisfaction. If doctors are expecting to tackle approximately three to six patient concerns during the visit, it is important that we as patients are setting that agenda. By focusing the doctor on the issues we personally want addressed, we are more likely to leave the office with a plan in place that satisfies everyone.

Any treatment plan needs to be a collaborative effort. When you, the patient, are involved in your treatment, you'll be more satisfied and have a better health outcome. According to the Arthritis Foundation, your doctor will also be less likely to generate unnecessary tests and referrals.

The status quo in medicine is neither acceptable nor sustainable. In the beginning of the book, I introduced the present epidemic of chronic diseases more recently called noncommunicable diseases. We all have them or know someone who does. But the problem is that modern medicine as currently practiced has few answers for the epidemic of these chronic diseases. If it did, the prevalence of these diseases would be noticeably reduced. People would be cured. Instead, more and more medical treatments and drugs are required by more and more patients.

We will continue to be thwarted by natural disasters within our own bodies and the continuing NCD epidemic until we treat humans as an ecological system to be personally managed from cradle to grave and across generations. The new biology considers the thousands of species that are a central part of who we are. It considers the gene switches (epigenetics) as a key element in our developmental programming, health, and well-being. It is time for this new biology to move precision medicine toward the more useful path of superorganism medicine. Such changes bring some uncertainties but also a completely new strategy for protecting human health, treating diseases, and ensuring human well-being. You as the patient are a superorganism, and any medical strategy should cover all of you, including your microbiome.

7

THE IMMUNE SYSTEM GONE WRONG

Your immune system is a junkyard dog. It can be your best friend, the ultimate protector of your health, and a lifelong partner to the end, supporting every tissue in your body, or it can make you sick and sometimes kill you. That may come as a bit of a shock.

We tend to think that the immune system is the only thing standing between you and certain death from infections. That is true as long as your immune system is functioning well and is well trained. When it is not, it can easily kill you. My own research on and teaching about the immune system beginning in the 1970s is not so much the happy story about the immune system and all the great work it does. Mine is the flip side. Mine is the story of natural disasters caused by an immune system gone so far rogue that it is literally blowing up the body. It can look much like a scene from a Bruce Willis, Arnold Schwarzenegger, or Dwayne "The Rock" Johnson action movie or one of your favorite disaster films (*The Perfect Storm*, *Independence Day*, *San Andreas*, *Titanic* . . .). Except in this case the location is your body.

"When Things Go Wrong" is the name of the block of lectures I give in Cornell's Basic Immunology course. Most students taking the Basic Immunology course do so as juniors or seniors but with only a cursory prior exposure to the immune system. You might say that my scientific career gig within immunology has two basic areas of focus: (1) what goes wrong with the immune system and how that leads to disease and, of course, (2) how to keep people out of harm's way and off a disease-filled path in life. As a result, my lectures cover allergic and autoimmune diseases with a pinch of inflammatory diseases and conditions (e.g., obesity, some forms of heart disease, some forms of cancer, and depression) thrown in for good measure. Some of the diseases can be a type of personal natural disaster for many people, forcing their way of life to change dramatically.

Natural disasters come in many forms. During my time with the Cornell Center for the Environment, I had the pleasure of working with a distinguished Cornell engineering professor, Walter Lynn, whose expertise was in global natural disasters. Whenever there was an earthquake, volcanic eruption, tsunami, flood, or massive forest fire, he was usually on a plane to go lend his much-needed expertise. Among many leadership appointments, he served as chairman of the board on natural disasters for the US National Research Council. I have been hoping that a bit of Walter Lynn rubbed off on me as I seem to be dealing with more and more natural disasters, particularly those located within our bodies.

For many students, my part of the immunology course is a bit of a harsh reality. I start the first lecture by polling the students, asking how many have family members (including themselves) or friends who face different categories of noncommunicable diseases, focusing on allergic and autoimmune diseases and conditions. There are very few hands unraised. I do the same thing when I guest lecture in one of Cornell's largest freshman-level biology courses. The response is the same. It is the same with the veterinary students I teach. As students look around these classrooms at the raised hands, they become increasingly aware that these diseases touch virtually every student at Cornell in some way. They have grown to become such a part of our societal fabric that until we tally up exactly who carries these diseases and conditions and the toll they take on our lives and those of friends and families, we are almost numb to their increasing existence.

Part of the reason that these diseases can travel under the radar is that we tend to use a name-and-divide strategy within medicine that can make it difficult to see the forest for the myriad of trees. New diseases and conditions are given various levels of official designations almost every day. Just look at the propagation of autoimmune diseases and conditions between 1970 and today. As often told by Noel Rose, Johns Hopkins professor and famed autoimmune researcher, his early days were spent with barely double-digit numbers of recognized autoimmune conditions. Today there are many more than one hundred, and the number is growing. This is reflected in new drug development to treat all of these new autoimmune conditions, which disproportionately affect women more than men.

Similarly, if you look at the neurobehavioral arena, a good way to track the growth in the number of different conditions is to consult each new version of the psychological manual for diseases and conditions called the *Diagnostic and Statistical Manual of Mental Disorders* (*DSM*). With each updated version, the *DSM* grows with new entries.

Of course, there are reasons why. A contributing factor to this growth is that we find more ways to partition dysfunctions within the human body. First, our detailed knowledge of physiological systems and organs, along with improved ways of imaging and analyzing their functions, allows for greater distinctions to be made. The result is that one prior disease may be divided into two new distinct diseases. Second, new drugs can be developed for each new officially named disease. Therefore, it is in the interest of the drug companies to eliminate gray areas around diseases and conditions and have new diseases defined whenever possible. A potential new drug that did not reach a level of efficacy for a broad disease may be more useful if the diseases are refined. If you doubt this, just look at the expansion of neurobehavioral conditions and the growth of prescription drugs administered to address childhood behaviors that were unnamed one or two decades ago. Drugs are prescribed based on their government-regulated, label-approved use connected to physician-diagnosed diseases and conditions. With more diseases, there are more possible opportunities for an existing or new drug to be used. This is at least partly why the list of diseases and conditions and their acronyms increases each year.

I used a story about a real peanut allergy in Part One of this book. Let us return to that topic now to illustrate exactly how far and how fast public health has deteriorated due to a dysfunctional immune system and NCDs. Probably the two most influential people to raise the visibility of peanuts as a healthy and useful alternative crop are George Washington Carver, famed agronomist and inventor, and President Jimmy Carter, the former peanut farmer from Georgia. Carver was instrumental in helping poor farmers find a sustainable farming future using alternatives to cotton such as peanuts. At the same time he researched and developed many new uses for peanuts that helped to expand the demand and markets for the increased crop production. This was a win-win for agriculture in the southern states and helped to produce greater yields for what was viewed as a healthy food.

For anyone alive in the 1970s, Jimmy Carter's emergence on the US political landscape, first as the governor of Georgia and later as president, was a surprise not only because of his relatively short duration of political experience but also since his profession was peanut farming rather than law. It was also a boon to the peanut industry as more and more people had their attention drawn to this crop and its uses in a wide variety of foods and nonfood products (e.g., cosmetics). In fact, Carter's grassroots campaigners were nicknamed the Peanut Brigade, and they used such visuals as peanut-shaped campaign clasps with Carter's name on the front to help convert his campaign push into election and boarded buses called the Peanut Special for a subsequent Peanut Inauguration Parade, for which commemorative plates exist. In the Carter era of the late 1970s, consumption of peanuts in schools soared. How quickly things have changed.

Peanuts are being banned from some schools and/or nut-free zones established, and airlines are deliberating whether to ban them from flights. Anaphylaxis at 30,000 feet is not a good thing. The risk of by-stander exposure is becoming too great for peanuts to appear in close quarters. This food is on a path where it might be enjoyed only in the confines of your own home, and even then with warnings or consent statements needed for visitors.

Peanut allergy is the poster child for what has gone wrong with our bodies starting with the microbiome and the immune system. How did peanut consumption become eerily similar to cigarette smoking in terms of ever-increasing restrictions on when and in what specific spaces the product can be used? Would Jimmy Carter dare call his grass-roots campaign the Peanut Brigade today?

Clearly, humans have changed since even the 1970s. If you look at food allergies and intolerances, we are seeing prevalences and intensities of adverse reactions that are both alarming and a reflection of our body's natural present disasters. Food certainly has changed as well (as discussed in Chapter 9), but our altered love affair with peanuts in only three to four decades reflects a major environmental shift in us. With 70 percent of our immune system residing in the gut, it makes biological sense that the front line of determining tolerance versus allergy to foods like peanuts involves our gut microbes and what happens in our gut.

So stepping away from President Carter and the 1970s and back to that present-day undergraduate immunology course, the present-day students, via their extended families and their friends, are touched by the very diseases I cover in my lectures. It's not just peanut allergy or the larger category of food allergies. It is also asthma, type 1 diabetes, celiac disease, multiple sclerosis, autism spectrum disorders, autoimmune thyroiditis, lupus, arthritis, atopic dermatitis, Crohn's disease, ulcerative colitis, allergic rhinitis (hay fever), psoriasis, and on and on. When something goes wrong with the immune system, the result is disease, which can show up in any tissue at any age. How is that possible?

Well, a little-known secret of biology and immunology is that immune cells are residents in virtually all of your tissues. They are placed there early in development and become so different from their cellular kin in other tissues that they are often given their own names. For example, what do microglia cells in the brain, Kupffer cells in the liver, and skin macrophages have in common? They are all macrophages. But their appearance and features are quite different. They morph while residing in the different tissues to take on their special characteristics. These resident immune cells exert significant control over what happens in those tissues. Ever wonder how you are able to see skin tattoos and why they last a very long time? It is because you are actually staining skin macrophages—they are the ones who take on the task of keeping the dye from reaching your internal tissues.

It is important to keep in mind that when these resident immune cells are happy and functioning well, the tissue usually functions well. But when the resident immune cells of a tissue are dysfunctional, that tissue is likely headed to a future of pathology and disease.

The common factors are that the immune system methodically attacks things it should not be attacking or that the response is completely uncontrolled. The response is often either completely over the top in intensity and/or never-ending instead of temporary. All of those types of inappropriate responses by the immune system result in pernicious damage to tissues and organs.

How can the immune system of the 1970s and before have been OK with peanuts, OK with the thyroid, OK with the skin and gut, but the immune system of the twenty-first century is misfiring anywhere and everywhere? The answers lie in how the immune system is trained in early life. Our first-genome human mammalian genetics today are not that different than those of the 1970s or even the 1920s. There are some genetic variants there that can affect the risk of developing an allergy or multiple allergies (called atopy). They contribute a small portion to the risk of immune problems. But what changed massively between forty and a hundred years ago is the early-life experience of our immune system. It is one reason some allergists are suggesting consumption of peanuts during pregnancy and early infancy and also having a dog in the house during the pregnancy and early infancy if you ever intend to have one. What your baby's immune system sees during this training period is what counts. But as we will see, potential allergens are only part of the story. There is much more at stake affecting whether the immune system will misfire during life and produce damage and disease, and that involves the baby's microbiome. Missed training of the immune system, which occurs all too often in today's babies, is a program for a different kind of natural disaster.

How the Immune System Develops

In a prior book coauthored with my wife in 2008–09 and published in early 2010, we described how the immune system develops and what things were known to affect its development. Development of the immune system includes two key aspects: maturation of the different cells of the immune system and education of the immune system in how, when, and where it should react. Because the immune system is widely dispersed in the body and very complex in terms of having many different specialized cells, it is very sensitive to developmental disruption. However, not every minute of prenatal or postnatal development is equally important for development of the immune system. There are different periods when major developmental events are occurring or immune school is in session, and there are other periods when immune development is comparatively quiet or school is on vacation. The most sensitive periods of immune development are called critical windows of immune vulnerability.

During these critical windows of immune vulnerability, the immune system is hypersusceptible for programming later-life dysfunctional responses and diseases. Disruption then means the immune system becomes unbalanced in cell populations and/or fails to learn how it should respond when challenged. Events disrupting immune development can include environmental exposures to chemicals or drugs, or intense or persistent maternal or infant stress. If a maturational step is missed or the education of the immune system is interrupted, the entire system of distinguishing friend from foe can go very wrong. An improperly trained immune system is almost guaranteed to eventually produce disease. Some combination of three things is likely: (1) failure to react to a real threat, (2) reacting to a threat with the wrong kind of defense, and/or (3) attacking our own tissues. Immune-related disease is often life-threatening.

While many environmental factors affecting immune education were laid out and even prioritized in the book I coauthored in 2009, the significance of the microbiome was not evident. Its role was only beginning to emerge at that time. Now in 2015 the picture of immune education during developmental windows would look quite different precisely because of the impact of the microbiome. That is how fast the biology of immune development and its impact on health is changing.

The impact of the microbiome on immune education is so critical and so all-encompassing that it is key to protecting the immune system of children. It is not just another environmental factor. The ramifications of environmental exposures and immune programming must now be seen through the lens of our microbes and recognition of you as a superorganism.

Remember, the microbiome is your body's ultimate gatekeeper. However, sitting between the microbiome and most of the rest of your body is the immune system, which is the next line of communication with the world outside of you. Yes, there are epithelial cells and linings in several of the sites such as gut and airways, but once you get past the skin, the immune cells are always just on the other side of the barrier. They are your mammalian cell greeting party, the welcome wagon as it were.

Not surprisingly, some of the most primitive, least sophisticated immune cells (representing your innate or natural immune system, as op-posed to your acquired immune system) are the very cells in closest and most frequent contact with the microbes inhabiting each portal of your body (i.e., gut, skin, airways, urogenital region). They have a temper, are highly mobile, and you don't want to upset them. In fact, consider them as the very core of that junkyard dog. They need careful, early-life train-ing, or they can be unpredictable and dangerous. Close friendship and contact with a complete microbiome is very important in the immune system's education. This happens both from physical interactions (al-most the equivalent of hugs or cuddles) and from chemical signals that are present in the metabolites of the microbiome. If the immune cells do not see enough of the microbes and get the right microbial signals in their early, formative period of education (shortly after birth), the im-mune system goes very wrong. It is almost a matter of when, not if, prob-lematic immune responses will happen. The BFFs need to be together throughout infancy. In the end, your immune system is the arbiter of what gets attacked and what is tolerated. It largely controls your risk for having allergies, autoimmune disease, and/or a host of other inflamma-tory diseases and cancers.

The innate immune cells, your most primitive, are found in the most primitive and ancient organisms. There are some organisms that do not have the types of responses immunologists call acquired or adaptive immunity, essentially immune responses to vaccines. They lack those necessary immune cells. But if they have any immune cells at all, they have innate immune cells such as some form of macrophage. This is not just coincidence. If microbes have been living with and communicating with host defenses of vertebrate and invertebrate animals since the start, then macrophages are going to be in all those animals, even if more sophisticated immune cells (e.g., certain types of lymphocytes) are missing.

A great deal is known about invertebrate immunology from the foun-dational work of Edwin Cooper of UCLA and his trainees. In invertebrates such as earthworms, there are innate immune defenses but not acquired immunity as we know it in mammals. Even amoebae have macrophage-like activity and can attack bacteria using macrophage-based functions when needed.

To emphasize the point about the BFF relationship between the microbiome and innate immune cells, Czech researchers recently found differences in the microbial-driven innate immune responses of two closely related species of earthworms that live in quite different natural composts. The species that lives in a manure-based compost that is rich with pathogens and needs robust immunological defenses had a much higher level of innate immune activity than that of the closely related species of earthworm living in forest mulch compost (with fewer pathogens). The scientists concluded that the microbial environment was the primary driver of the status of what were quite comparable innate immune systems. The fact that the immune system is primitive should not discount its importance within humans.

The Misunderstood Immune System

The common view of the past decades is that the immune system establishes a fortress wall and defends our mammalian body against invading microbes. That is what I was taught in college. The other thing I was taught in college was that the immune system (1) resides in a limited number of body sites that are specialized lymphoid organs (thymus, spleen, bone marrow), (2) travels in blood and lymph, and also (3) samples our portals of entry for exposure to invading bacteria and viruses. There was virtually no mention of the gut as the main location of immune cells, despite the fact it has a majority of our immune cells, or the fact that virtually every tissue and organ of our body has its own mini immune system permanently residing there. This led to some misconceptions about what the prime directive of the immune system actually is. It is not just sitting in all our tissues to sample for the first invasion of microbes—after all, the liver and brain are not the first sites where microbes gain entrance to infect. Instead, the immune system is sitting in the liver, brain, and other tissues and organs to control their integrity and to help control the balance of function in our specialized tissues.

Ironically, the groups of immune cells living within our different organs (brain, liver, kidney) are so radically different in appearance and some properties that scientists weren't even certain they were immune cells when I went to college. But these highly specialized immune cells are in our specialized tissues for a purpose other than just microbe hunting. We now know that microbes aren't the only threat. Internally developed cancers are something the immune system must deal with as well. Plus, the immune system clears us of all dead and dying cells much like an overnight building custodian who wants to disrupt normal operations as little as possible.

So, in addition to pathogen hunting, the immune system is analogous to an environmental and security control system in an office building. It is an integral part of virtually all of your organs, ensuring that the conditions are met and maintained for effective organ function. When the immune residents in your organs function well, your organs are likely to do so as well. But if those immune cells go rogue, your organs are in serious jeopardy. Inappropriate immune responses within the organ cause organ damage, loss of organ function, and the increased chance of cancer involving that organ or tissue. The resident immune cells are also likely to signal for help from external immune cells, which come rushing into the organs, attaching to our normal cells. This can cause organs like the thyroid and pancreas to become more of an immune organ than an endocrine organ in a comparatively short time (e.g., autoimmune thyroiditis, diabetes).

Why would your immune cells do this? Why would they divert from protecting the integrity of our organs and tissues to inflicting harm on us? There are several reasons why this could happen. But I will argue here that the most significant reason for immune-inflicted noncommunicable diseases is the loss of a higher order of self-integrity involving our microbiome. If the immune system matures in an environment where we do not self-complete and are missing our intended microbiome, our immune system gets programmed for haphazard, inappropriate responses. It is then only a matter of when and in what tissue disease will show up. Will it be in our brain, causing neurobehavioral and neurodegenerative issues; our liver, causing metabolic issues; the gut, causing digestive-inflammatory issues; our endocrine organs, causing hormone/metabolic problems; our bones, causing osteoporosis; our mouth, causing dental cavities; our blood vessels, causing cardiovascular disease; or in any of these locations, causing cancer?

Remember those macrophages I discussed earlier that reside in all our tissues. They appear to be able to morph into different forms, are given different names, and can operationally control tissue function, plus potentially destroy the tissue if they so choose. I have told my students half jokingly that macrophages rule the world, and if only we knew how to control macrophages, it would be a better world. Of course now we do know how to control macrophages . . . It is through the microbiome.

Asbestos and the Junkyard Dog

Most people have probably heard of asbestos, even if they are a little fuzzy about why. They may have seen danger signs in hallways or on the outside of old buildings announcing asbestos remediation areas or have seen one of countless TV ads from law firms about class action suits, asking if you or a loved one has been diagnosed with mesothelioma caused by exposure to asbestos. There is even a 900-plus-page book intended as a guide for lawyers concerning asbestos health litigations. Some people may have come across a recent article in *Scientific Ameri-*

can asking parents if their child may be coloring with crayons containing asbestos, including comments on the dangers of such exposures from a leader in children's health protection, Dr. Philip Landrigan of Mount Sinai's Kravis Children's Hospital in New York City. I have memories of specific asbestos products from my younger days. In fact, for a period of time, you could walk into any well-equipped research laboratory and find pairs of asbestos-lined gloves. They were the gold standard for handling hot lab ware from sterilization ovens. For the majority of the twentieth century the health risk was simply not known. It was thought that they protected lab workers without presenting a significant health risk.

There are other parallels right in the research laboratories. The immune-toxic, cancer-causing chemical benzene was used in twentieth-century chemistry labs as the go-to solvent for cleaning glassware. Only later did the health risks become apparent. Benzene went from being a common type of liquid detergent to something stored and used under the most highly contained lab conditions and with workers fully protected. In mid-twentieth-century homes carbon tetrachloride was a common cleaning aid ready at hand when your guest spilled a drink on your beautiful new sofa or your pets had an accident. But no more. Its use in consumer products was banned in the 1970s.

So asbestos was not alone in being almost a miracle material in widespread use in the twentieth century that was later recognized and regulated as a significant environmental health hazard once its action on humans was fully understood. But exactly what is asbestos and how does it affect our health via the immune system?

Asbestos is a series of naturally occurring mineral fibers that can be separated into thin threads. It is mined much like the metals gold and uranium. It was extensively used in construction materials and car parts and could be found in some gardening products and even some products designed for children. Among the hot spots for asbestos exposure were mining operations near Libby, Montana, for the production of vermiculite. Between the late 1970s and today, there have been a series of bans on asbestos use in different products, and this has led to a significant reduction in its annual production in the United States. What does asbestos do that makes it so dangerous? Actually, asbestos is primarily toxic and carcinogenic via the immune system.

By themselves, the fibers don't do a great deal. The problem is they don't easily degrade or go away. That is a problem when they end up on innate immune sentry cells like macrophages in your lungs (called alveolar macrophages). These macrophages, along with cells of the airway linings, engulf and accumulate the asbestos fibers. But the macrophages can't digest them. So they respond in several ways. They accumulate near the airway borders and mount a massive, never-ending inflammatory response featuring damaging oxygen radicals that overwhelms your antioxidative defenses. If this goes on long enough, cancer can and does result. Mesothelioma is one of the predictable outcomes.

At the same time, macrophages and other innate immune cells lead a repair effort as lung tissue damage builds up in the area. But the repair effort has two critical features: First, the repair uses biological materials that fill the lung space but don't replace lost function. In other words, the repair does not help you to transfer oxygen to the blood as is needed. Also, the repair blunts anti-tumor responses by other immune cells. Macrophage changes in the lung with asbestos also allow for autoimmunity and for cancers to survive better. This is especially bad news since, after the decades of oxidative damage, the risk of tumors forming in the lungs is high. The innate immune attack against asbestos led by macrophages can result in several forms of lung disease. In the end, because these cells cannot digest the asbestos, they lash out, and the lungs pay the price.

This is precisely the type of scenario that happens over and over in our bodies when the immune system does not cope well with environmental exposures. Only, instead of asbestos, it could be exposure to truly innocuous factors that cause dysfunctional or untrained immune cells to produce self-damaging, inappropriate, inflammatory reactions in the brain (neurodegeneration), the reproductive tract (sterility), the pancreas (diabetes), the gut (inflammatory bowel disease), the lungs (asthma), the skin (psoriasis), the heart (myocarditis), the skeleton (osteoporosis), or the liver (several forms of hepatitis).

Finally, a lingering question is whether there is a role of the lung microbiome in asbestos-related lung disease. Researchers recently suggested that one of the factors affecting the risk of asbestos-related cancer is whether the fibers penetrate the lining of the airways and reach the alveolar macrophages. They hypothesize that one of the determining factors in penetrating the lining may be whether lung microbes secrete proteins that punch holes in the epithelial cells providing the lining border. That would confirm the idea that the microbiome is our gatekeeper, controlling what actually reaches our internal cells. It remains to be determined if this is a major factor in internal asbestos concentration.

Training the Immune System

A wonderful and health-promoting discovery uncovered between the years 1979 and 1984 may, in the long run, end up being viewed as a tipping point for the fallout over the twentieth-century war on microbes and overuse of antibiotics. Barry Marshall, the Australian physician, and his collaborator Robin Warren reported the link between a spiral-shaped bacterium, only found in humans, named *Helicobacter pylori* (*H. pylori*), infections of the stomach, and peptic ulcers. Their report was first published as letters in 1983 with a full report in 1984 published in the medical journal *Lancet*. Prior to that time a diet of spicy foods and mammalian genetics were thought to be the major factors determining who got ulcers and gastric cancer. For their discovery, Marshall and Warren were jointly awarded the 2005 Nobel Prize in Physiology or Medicine. The solution seemed simple enough: Eradicate *H. pylori* with massive antibiotic therapies anywhere and everywhere it showed up. That was in keeping with the widely held twentieth-century view that the only good bacteria are dead bacteria should they dare to come into our bodies.

Yet at the same time as the Marshall and Warren discovery, other researchers held a slightly different view. In his book *Missing Microbes*, Martin Blaser describes his lengthy, contrasting research into the health-promoting activities of *H. pylori* as a stomach resident. How could the same information be taken to different conclusions? It happens because the whole human is an ecological system, as discussed in Part One of this book. Not every cohabiting species within or on us is innocent of causing potential harm, nor are potential pathogens always without any redeeming value to us as a superorganism. It can be situational. *H. pylori* has been associated with humans for thousands of years, having been found in mummies from northern Mexico dating to before the arrival of Columbus in the New World. Balance is needed, combined with broader understanding of who and what we are. It turns out that *H. pylori* and its multiple effects probably fit into an immunological idea known as the hygiene hypothesis, first described in 1996 by David Strachan of the UK.

When it comes to microbes, just as with environmental chemicals, here's a prescription for good health: the right place (specific body location), the right amount (dose), the right time (developmental, menstrual cycle, or circadian cycle timing), and compatible with our mammalian self. The flip side is equally a path toward pathology and disease. The wrong place, the wrong amount, the wrong time, or incompatible with our mammalian self often causes significant health problems.

Why might you want some *H. pylori,* which in some circumstances can produce peptic ulcers or stomach cancer? Because purging the body of *H. pylori* is equally associated with other NCDs, and we have a good idea about the mechanism of how this happens. It turns out that the persistent presence of *H. pylori* helps the immune system become more tolerant and reduces the risk of asthma, allergies, and inflammatory diseases such as inflammatory bowel disease (IBD), beginning with immune cells called dendritic cells that sample the environment. Partly through *H. pylori*'s effects on dendritic cells, there is an increase in the maturation and numbers of regulatory T cells called natural Tregs, and these are critical to avoiding the inflammation that supports numerous NCDs such as asthma, allergies, and IBD.

This example illustrates that we need microbes in place that will educate and train our developing immune system, or we are likely to face a plethora of different inflammation-driven NCDs. If it is not *H. pylori*, then some microbial equivalent needs to be in place in our microbiome to ensure that our immune system does not go rogue and accepts both our own tissues and the harmless things in our environment.

Early Microbial-Driven Education to Avoid Disease

A prime example of the importance of early education of the immune system by the microbiome comes from the collaborative immunology and gastroenterology efforts of Dennis Kasper and Richard Blumberg and their laboratories at Harvard Medical School. These groups used the C57 Black 6 strain of mice, which is a standard research model for immunology, to investigate the effects of commensal bacteria and their metabolites on early immune maturation and susceptibility to later-life noncommunicable disease. In this case, the disease was colitis, which is similar to ulcerative colitis in humans (one of the two parts of inflammatory bowel disease). In this strain of mice, the lack of commensal bacteria makes the mice highly susceptible to colitis when they are exposed later in life to oxazolone. The immunological mechanism for this colitis is well described, with specific immune cell populations and immune hormones leading the way to produce disease. The dysfunctional immune process in mice appears to be similar, if not identical, to that which produces human ulcerative colitis.

What the two research groups at Harvard did that was so intriguing is they asked four important questions about the microbiome, immune dysfunction, and susceptibility to colitis. Their first question was whether a single commensal gut bacterium could protect against later-life colitis. The answer was yes, and the bacterium that provided resistance to the disease was *Bacteroides fragilis*. This bacterium is rod-shaped, does not need oxygen to grow, and is usually one of the immune system's friends as long as it stays put in its region of the gut.

The Harvard researchers then went a step further and asked if there was a critical developmental window when the bacterium had to be in the newborn mouse to produce resistance to colitis. The answer was it had to be there by one week of age. After that, the addition of *B. fragilis* could not prevent colitis.

Next, they asked how the bacterium made the mice resistant to disease in the newborn. The answer was it blunted the proliferation of a specific population of gut-innate immune cells called invariant natural killer T cells (iNKT cells). In the absence of the bacterium, these cells had a burst of proliferation in the infant mouse, and that burst was what made the mice susceptible to colitis for the rest of their lives. If the bacterium was present in the gut, the burst was significantly lessened and the mice were resistant to colitis for the rest of their lives. It was a remarkable finding about the importance of an early developmental window for the immune system and the need for the microbiome to be in place to avoid later-life disease.

Finally, the researchers asked whether the entire bacterium was needed or if a metabolite of the bacterium could produce the same beneficial effect on the developing immune system. They found that a particular type of lipid made by *B. fragilis*, given at just the right developmental window, could suppress the amount of iNKT cell proliferation and make the mice resistant to colitis. This is one gut bacterium, one immune change, and one NCD. Imagine the opportunities to reduce the prevalence of NCDs when the entire microbiome can be effectively managed to support the best possible cultivation of your immune system.

It is obvious perhaps, but this fact about death from infection needs to be emphasized. Unless the infection causes immediate failure in vital organs or blood vessels, such as with the Ebola virus, risk of death usually comes down to how the immune system responds to the infection. The Spanish flu pandemic of 1918–19 took an estimated 21.5 million lives globally, including approximately 675,000 in the United States. But many people were infected and survived. In fact, if you look at who died and who got infected but didn't die during the pandemic, it came down to which individuals mounted responses that, in attempting to kill the virus in the lungs, led to complications that compromised lung function. It is thought that they mobilized an unhelpful, aberrant inflammatory response. The immune system can save your life, or it can kill you. It just depends. In the case of that flu, it appears that an overzealous, unrelenting immune attempt to purge the body of the virus led to many deaths, particularly among healthy young adults.

The same thing can and has happened with bacterial infections. Wiping out bacteria with the use of antibiotics does not mean a patient automatically survives. Some pathogenic bacteria carry what are called toxins. These toxins are chemicals in the bacteria's outer cell makeup that cause immune cells such as macrophages to go just a little crazy and start shooting first and asking questions later. This is a part of what we usually refer to as the inflammatory response. The inflammatory response is a good thing if it is in the right place, of the right kind, at the right level, and ends when it is no longer needed. Anything other than that is a problem.

Bacteria are usually not a huge problem if the bacteria are few in numbers and localized to one place in the body. But even if the bacteria die, such as after exposure to antibiotics, their outer layers or shells carrying the toxins must be cleared from the body. This is a job for macrophages and their friends. There are two major categories of bacteria: gram-positive and gram-negative. For our purposes the important thing to know is that each carry different sets of toxins. Gram-negative bacteria are powerful direct activators of innate immune cells (e.g., cells such as macrophages and neutrophils that are capable of a generic type of immediate response to pathogens). Some gram-positive bacteria have toxins that cause massive T lymphocyte (thymus-derived lymphocyte) activation, producing a storm of immune hormones known as cytokines. This cytokine storm, in turn, activates macrophages for destruction. An example of this in humans is toxic shock syndrome.

If the toxins have actually made it into the blood at high enough levels, innate immune cells in the blood go crazy and start shooting there. That is never a good idea. It destroys the blood vessels and produces what is call septic shock. This is very serious, and patients often have only minutes to receive treatment or they will die. Sometimes the toxin levels are so high it is difficult to save the patient. That is why physicians like to kill bacteria slowly with antibiotics (over one to two weeks) and not all at once. Here again the dead bacteria are not killing the patient. The bacteria are already dead. Innate immune cells responding to the perceived threat are killing the patient. Does this actually lead to death? Yes. Your host defense system can unintentionally destroy itself. In fact, Dr. Kevin Tracey has catalogued numerous examples of why what are usually nonfatal infections can lead to immune-inflicted death in his 2006 book *Fatal Sequence: The Killer Within.*

An improperly educated, dysfunctional, or out-of-control immune system can make you very sick and kill you. You need the immune system to develop appropriately, to function in a controlled manner, to function in balance, and to recognize and respond to actual threats but tolerate and leave alone your healthy cells and tissues as well as innocuous environmental factors (e.g., foods and allergens). The best way to ensure this happens is to connect the immune system with a healthy microbiome from birth to have the immune system properly educated and brought into balance. We now know that protecting our microbiome has this essential purpose.

8

PATTERNS OF DISEASE

P atterns are everywhere we look. They are not just a part of the fabric of human clothing, raising the question of whether a striped shirt goes with plaid pants. They are a part of the fabric of the entire universe. Galaxies and solar systems are organized into patterns. There are patterns for how our communities, cities, and languages are organized and used. There are patterns within plants, among the plants in a forest, and among the microbes inside and on us. Finally, there are patterns to human disease. Patterns have always fascinated me because we can learn more about how things actually work if we understand more about the larger pattern they fit into. You can stare at a single piece from a jigsaw puzzle by itself and may envision an almost limitless number of potential uses for that piece. It could fit into a thousand different puzzles. But once you see the whole puzzle it *does* fit into, the role of the single piece becomes crystal clear. Suddenly, you know a lot more about that single puzzle piece. That is how I see human disease. I will describe how seeing the whole puzzle or larger pattern for human diseases is very important when it comes to tackling those diseases.

When diseases are given distinct names and specific medical codes, paths for new research funding, advocacy groups, and treatment options open. There is nothing deeply wrong with this approach. But the disease-naming process, in which minute distinctions designate the differences that separate one NCD from another, can shift our focus away from the common ground these diseases have. It seems that no one is looking for similarities and the common denominator between diseases that could be used to fight them all.

It can be very constructive to look for patterns of similarity that link things together. It is an imperative that has driven much good science over the centuries. And when it comes to NCDs, the ailments are far more similar than different.

No matter which organs or tissues are involved, NCDs are fundamentally connected in ways that are only now being understood. First, they are connected in the likelihood they will show up together in the same person. In other words, the person will get one NCD diagnosis followed later by a second and even a third. One example would be obesity followed by diabetes.

When they show up together, they are said to be comorbid diseases, or to demonstrate comorbidity. We may not always know exactly how they are connected, but they seem to always have the same traveling companions. I will talk a lot about comorbid NCDs and the comorbidity of NCDs. They are a little like cockroaches. If you see one out in the open, you can be pretty sure there are tens to hundreds more just waiting to make their appearance. NCDs are connected biologically, epigenetically, metabolically, and, most importantly, microbiologically. Of course the microbiological connection of NCDs is through our microbiome and is what can promote the other connections as well. The exciting aspect about these interconnections is that they will allow us to attack NCDs in groups rather than singly. Considering they are all part of the same epidemic and are interconnected in many other ways, it only makes sense to go after them as a group.

Here is an example of what I mean by the comorbidity of NCDs. In this case I am using obesity, one of the most well-known NCDs, as a starting point to show you what a pattern of interconnected NCDs looks like. Obesity is all too common, and we are in the middle of an obesity epidemic. In the United States in 2011–12, more than one-third of adults past the age of twenty were obese and more than two-thirds of the adult population were overweight. This is in contrast with the US population of the 1960s, where the adult obesity rate was about 13 percent, with the percentage of overweight adults still less than one-third of the population. Children are not immune to this obesity epidemic. In the most recent analysis, 20 percent of adolescent children were obese.

The following figure shows recent information on the NCDs that are connected to obesity. By looking at the diseases connected to obesity, you can get a better picture of the real lifelong impact of a single NCD like obesity. It is not just one disease or condition. It is the entrance into a pattern of multiple likely diseases.

Like many other NCDs, obesity is said to be a pro-inflammatory condition. Inflammation does not end when it should in obese individuals. This ongoing low-level inflammation is unhealthy and leads to many other health problems involving our immune system and different tissues. I use the phrase "unhealthy inflammation" because I am talking specifically about inflammation in our bodies that is wrong for the job at hand. It doesn't help us. In fact, it damages us. Most often unhealthy inflammation is wrong because, as in obesity, it doesn't stop when it should. Unending inflammation should resolve, but doesn't. It just keeps going and eventually will promote disease. Such chronic inflammation is thought to cause diseases like cancer. Sometimes inflammation is unhealthy because it is misdirected. It attacks the wrong targets, such as our own cells instead of pathogens.

Obese individuals are at a greater risk for myriad diseases, including, at a minimum, the thirty-two shown in the following figure (which includes twelve different types of cancer). These are essentially the fruits of the initial NCD, in this case obesity. Note how these thirty-two diseases fall into different medical categories (e.g., cancer, heart diseases, neurological disorders, endocrine and metabolic diseases, autoimmune and allergic diseases). Knowing about these interconnections can help prevent the additional diseases. Also, by recognizing the interconnections and commonalities among these NCDs, we are better able to seek comprehensive solutions rather than only using piecemeal, single-disease therapies.

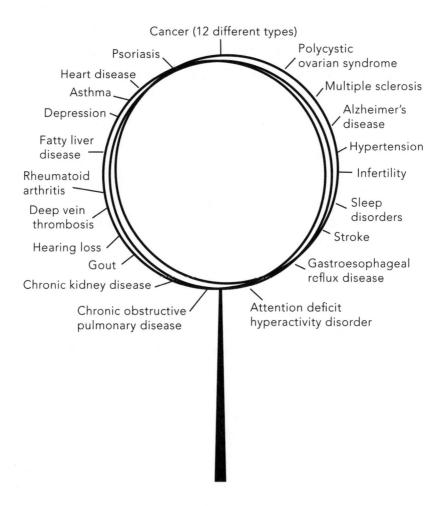

THE OBESITY TREE

A Pattern of Thirty-two Interlinked NCDs for Obesity

This is needed because recent preventative and therapeutic initiatives have not reduced the NCD epidemic. While the World Health Organization focused on smoking reduction, healthier eating, and increased exercise to solve aspects of the NCD problem, the recommendations have produced only modest outcomes at best.

It was 2008 when I became interested in the NCD epidemic, and I began working on uncovering the patterns among the NCDs that might help design better, more holistic strategies for disease prevention as well as therapy. With the help of three colleagues—Judy Zelikoff at NYU School of Medicine, Dori Germolec at NIEHS, and Jamie DeWitt at the East Carolina School of Medicine—I began to examine how NCDs were connected. We described several different interlinking patterns of NCDs, much like the example I showed for obesity, and these were published in a series of papers in pediatric medicine and environmental health journals.

What we discovered was beyond anything we could have imagined. It came down to four essential elements or pillars of the NCD problem:

The Four Basic Pillars of NCDs
1. NCDs get programmed in early life (from before birth until around age four).
2. Uncontrolled inflammation maintains the disease state, with microbiome problems initiating and maintaining the unhealthy inflammation.
3. Having one NCD increases the chances of developing one or more further, specific NCDs.
4. Microbiome status affects the risk of NCDs and determines the effectiveness of drug therapies.

Let's take a deeper look at these four pillars.

1. Early-Life Programming for NCDs

As discussed in Chapter 5, our physiological systems get programmed very early in life. Importantly, this programming involves all our systems, especially the immune system, which in turn affects our risk of developing NCDs. In these early years our microbial partners help our systems mature and program our future health. This is why having a complete microbiome at birth is crucial for a healthy life.

One way to see this early programming is to look for key indicators that a disease template is already in place. Catching the indicators can point toward diseases before they even show up. Astonishingly, even adult-onset NCDs can be evident in the infant. My colleagues and I then realized that catching and treating the NCD early could prevent it from developing in middle age, when nothing further can be done beyond symptom management.

Take atherosclerosis, a form of cardiovascular disease, for example. Heart disease is one of the leading causes of death in the developed world. Atherosclerosis causes arteries to harden due to the formation of atherosclerotic plaques. The plaques consist of artery-filling, cocoon-like structures surrounding a rogue, fat-filled macrophage called a foam cell. Plaques change the properties of the entire artery as they accumulate. The arteries break; blood clots form; arteries become totally blocked. At this point the heart and brain are starved for blood and oxygen and a heart attack or a stroke results. Unfortunately, foam cells live a long time and the buildup can occur slowly. Though the fully developed disease is usually diagnosed in men around fifty and in women from sixty and up, the beginnings of the disease can be detected in the infant decades earlier. Indicators of chronic vascular inflammation in children, such as increased levels of C-reactive protein, oxidized lipids, pro-inflammatory cytokines, and endothelial dysfunction, are useful predictors of whether later-life atherosclerosis will occur.

If the microbiome is fully in place and inflammation remains optimally controlled, the disease can be avoided since it is totally reliant on unhealthy inflammation. Problems with the microbiome can cause unhealthy inflammation and aid development of atherosclerosis. But just as easily, correction of a problematic microbiome through use of probiotics is reported to reduce inflammation and also the risk of atherosclerosis.

2. Unhealthy Inflammation

Virtually every NCD has unhealthy inflammation with excessive oxidation at its core. Oxidation is a normal chemical reaction that utilizes oxygen. But some of the by-products of oxidation are free radicals that can damage our cells and tissues and alter our proteins and DNA. Free radicals of both oxygen and nitrogen are useful if you are trying to kill pathogens. They are harmful if our own cells encounter them. That is one of the reasons to eat a diet filled with antioxidants and sometimes take antioxidant supplements. We need to avoid oxidative damage of our own cells. But inflammation usually produces oxidation, and unhealthy inflammation produces so many free radicals that it is impossible to scavenge them all with antioxidants. When this happens and our tissues receive years of oxidative damage, they either lose tissue function or the tissues' cells become cancerous. Inflammation as nature intended is both necessary and useful. However, it needs to be:

1. appropriate for the task;
2. directed at the narrowest target possible (the pathogen, not an entire organ); and
3. resolved once the legitimate threat has passed.

Unfortunately, all too often inflammation begins and continues in a completely uncontrolled manner, like a California wildfire during a prolonged drought (or a junkyard dog).

When inflammation rages out of control for too long, the inflammation shifts from attacking a perceived pathogen to walling it off and isolating the perceived danger. The science fiction drama television series *Fringe* portrayed an analogous process, albeit on a larger scale. The "amber zones" in the show were nonfunctional, unusable, and uninhabitable regions. They were created to wall off mini black holes that were highly unstable and dangerous. They were called amber zones because they were sealed off using an amber-like hard plastic substance. The amber zones were posted with signs and encompassed major sectors of cities. Anything that accidentally entered the area had to be sealed off in amber as well. Much the same thing occurs in our bodies. Things like tuberculosis and asbestos exposure cause the immune system to initiate this type of walling-off response in the lungs, eventually creating one big nonfunctioning amber zone. After enough walling off within the lungs, we may not have enough function left to survive.

The big question about the interconnections between NCDs is how one thing, inflammation, could be at the heart of them all. To answer this, we have to look at the most basic type of immunity, innate immunity. Innate immunity is what your body develops naturally without the need for vaccinations or immunizations. Its most basic function is to send out scouts throughout the body that watch for harmful pathogens. When they are spotted, the scouts call in the specialists, which sweep through, attacking and eating up the pathogens and dead cells. When you get something like strep throat, this process begins immediately. The inflammation produced by this process is what makes your throat sore and your temperature rise. If your doctor prescribes an antibiotic, all the drug can do is reduce the numbers of strep-causing bacteria in your system, thereby making it easier for your immune system to kill the bacteria and return you to good health. That is an appropriate immune/inflammation response. You wouldn't want the attack to occur in healthy organs and tissues, though. You want it to happen just where the harmful bacteria have clustered and last only as long as necessary.

This same type of immune response has been found in lower, less sophisticated animals. Elie Metchnikoff, a biologist from Ukraine, received the 1908 Nobel Prize in Physiology or Medicine for his discovery of phagocytes and phagocytosis, one of the most basic parts of the immune response. We now know phagocytes as macrophages, and they inhabit virtually every organ and system in the body. Metchnikoff made his discovery while studying the digestive organs of bipinnaria starfish larvae. He introduced dye particles and wood splinters into the larvae and watched while previously unnamed, independently moving cells surrounded and engulfed these foreign elements. This observation made Metchnikoff realize that phagocytes (our macrophages) are our first line of defense against infection. These are the cells that provide innate immunity and help you fight off illnesses like strep throat.

So, how do macrophages, which are designed to protect us against infectious diseases, cause NCDs? They do this when they continue causing inflammation long after the illness is over, mistakenly attack the body's own organs and tissues as if they were harmful invaders, and attack harmless things from the environment (e.g., pollen, foods, etc.). Why would macrophages act like this when to do so is against their design? If they never learned what is harmful and what is safe, or what belongs to the self they reside in, they are lost. This can happen if a baby's immune system does not mature from its newborn state. It is important to recognize that the newborn's immune system is neither fully matured nor balanced as the baby enters the external world. Both of these immune changes must happen for the baby to lead a healthy life.

Why can't the baby be born with its immune system completely matured and ready to go? It is because the baby develops inside the mom. As the baby develops within the mom's protective womb, both the baby's and mom's immune systems need to be set up not to attack each other. If the womb environment permitted it, they could attack each other because they are not genetically identical. To ensure this attack does not happen, some of the mom's immune responses are dampened by the womb environment, and the baby's immune responses that would at-tack what is foreign to it wait to develop. The womb helps to create an environment that is biased in favor of allergic responses and has a defi-ciency in the immune responses against viruses and tumor cells. This immune-skewed, dampening-down environment that the baby is living in prenatally affects what is happening with the baby's immune system. If anti-tumor responses were not suppressed, the mom's immune system would view some of the father's proteins carried by the baby as a type of tumor. An anti-baby immune response by the mom would result in mis-carriage. This must change upon birth or the baby's immune system risks never becoming fully mature and balanced. If those changes don't happen in the newborn, the baby will have health issues such as allergic, autoimmune, and inflammatory NCDs.

A complete microbiome must be in place at or just following birth to help the baby's immune system finish maturing and to balance out those immune responses. Co-development of the microbiome alongside the baby's immune system is critical for that baby to have healthy and appropriate immune responses for the rest of its life.

3. NCDs Lead to More NCDs as You Age

According to recent reports, just fewer than half of all Americans will have two or more NCDs by age sixty-five. Once you are diagnosed with the first NCD, a vicious cycle begins. As you age, your human ecological system falls apart.

To measure the impact of the NCD epidemic, we can examine things such as causes of death and increased drug therapies. According to the US Centers for Disease Control and Prevention, the leading causes of death in the United States for 2013 were heart disease, cancer, lung diseases, acci-dents, stroke, Alzheimer's disease, and diabetes. As I lay out the tight con-nections among NCDs, watch for how many times cardiovascular problems show up as secondary diseases to an initial NCD diagnosis. Also, notice how often the tissue bearing the brunt of the initial NCD (often in children) and the inflammation associated with that disease is the target of later-life cancer. It happens all too frequently, and it is very predictable once you begin to look at the patterns within our ongoing NCD epidemic.

In 2010 Kathlyn Stone, a journalist specializing in scientific research who is well versed in clinical trials and regulatory policies, reported that nearly four billion medication prescriptions had been written in the United States alone that year. There were increases in all classes of drugs from 2005 to 2010. Worse, 90 percent of senior citizens and 58 percent of nonseniors reported regularly relying on a prescription medication. That parallels the data that say NCDs are normally lifelong conditions that lead to more NCDs as you age and require drug therapy for symptom management. Of the four billion prescriptions that were filled, the most common were statins, used to treat metabolic issues and cardiovascular diseases. They were followed by antidepressants, antidiabetic drugs, sleep medications, antihistamines, and drugs for respiratory conditions like asthma and chronic obstructive pulmonary disease (COPD). Just in looking at the medications, the prevalent NCDs of cardiovascular disease, depression, type 2 diabetes, insomnia, allergies, and asthma are dramatically apparent. These diseases are points on a web of connections between NCDs.

I have already shown you the pattern for obesity in the figure on page 126. But let's look at the interconnected patterns for a few other well-known NCDs: asthma, type 1 diabetes, celiac disease, and autism. There are now many other similar patterns we could examine for diseases and conditions such as food allergies, inflammatory bowel disease, Alzheimer's disease, heart disease, breast cancer, ADHD, and sleep problems, but the small group that follows will illustrate the point.

Asthma, a lung disease, is another well-characterized disease for comorbidities. Asthma itself features improperly controlled immune responses and inflammation in response to allergens and other conditions that are not actually harmful and do not require our immune system to respond. Because of the constant inflammation in the lungs, the lung tissues experience regular attacks from cell-damaging chemicals produced by immune cells. If this continues too long, the lungs can no longer hold up and cancer develops. While asthma can kill, the disease is often managed over the course of a lifetime. But to date, the management of asthma does not reduce the risk of later-life lung cancer. Asthma is not just connected to lung cancer but also to several allergic conditions, neurobehavioral alterations, olfactory disorders, and being overweight.

Type 1 diabetes is a juvenile-onset autoimmune disease of the pancreas. It is linked with a dizzying array of comorbid diseases that extends well beyond other autoimmune conditions. Diseases identified as comorbid for type 1 diabetes include autoimmune thyroiditis, celiac disease, Addison's disease, vitiligo, eating disorders, depression, anxiety, osteopenia, colitis, cardiovascular disease, epilepsy, psychiatric disorders, and obstructive lung disease. Adding to this list, a recent study from Australia catalogued the significant cancer associations with type 1 diabetes. For both sexes with type 1 diabetes, there is elevated risk of cancer in the pancreas, liver, esophagus, colon, and rectum. Additionally, for females only, there is additional elevated risk of cancer in the stomach, thyroid, brain, lungs, endometrium, and ovaries.

Celiac disease is a gastrointestinal autoimmune disease linked with gluten sensitivity that has been rising in prevalence as well as showing up in younger age groups compared with a decade ago. In keeping with other NCDs, the reported comorbidities for celiac disease mostly fall into the same category of disease (autoimmune): autoimmune hepatitis, autoimmune pericarditis, immune thrombocytopenic purpura, pancreatitis, peripheral neuropathies, psoriasis, rheumatoid arthritis, sarcoidosis, Sjögren's syndrome, and type 1 diabetes. However, there are many other diseases comorbid with celiac disease that are not autoimmune conditions and, therefore, might be unexpected since they fall into other medically designated categories. For example, depression is common in women with celiac disease. Other comorbid diseases and conditions include COPD, cardiovascular disease, hearing loss, restless leg syndrome, osteoporosis, eating disorders (women), risk of miscarriage (women), eosinophilic esophagitis, and cancer in the target tissue (small bowel adenocarcinoma). Given the extensive list of comorbid diseases, it is easy to understand why more aged patients are likely to be diagnosed with additional NCDs.

Autism spectrum disorders (ASD) can produce a heavy burden for individuals and their families. But as with virtually all NCDs, ASD has its own quite predicable set of additional diseases that are more likely to occur in the individual with ASD than in the general population. Some of these are neurological conditions, but many are not. For example, in girls with ASD there is a form of epilepsy that is resistant to treatment. Gastrointestinal disorders such as food intolerances are more common in both sexes with ASD than in the overall population. Sleep disorders are another common condition that is co-occurring with ASD in children. This is not surprising, as sleep disorders and/or depression show up with a majority of NCDs. A disease connected to an immune disorder, mastocytosis, a hyperreactivity of mast cells (an allergy-related immune cell type), is rare in the general population but much more common among ASD children. A recent study of adults with autism found that they carry a heavier burden of multiple NCDs, including hypertension, diabetes, stroke, Parkinson's disease, sleep disorders, depression, schizophrenia, and bipolar disorder.

There is another way to look at the disease interconnections and that would be to ask, what do they all have in common? The answer is twofold. The patterns of interconnected NCDs feature the previously mentioned unhealthy inflammation as well as a major depressive disorder usually called, simply, depression. At first glance, depression and these other disorders might not seem to have anything to do with the status of the microbiome and the immune system, but suspend your disbelief.

Depression unlocks the puzzle of these connections. The microbiome exerts exquisite control over the immune hormone and inflammation problems that lead to depression. The list of ailments that are comorbid with depression include asthma, type 1 diabetes, type 2 diabetes, multiple sclerosis, cardiovascular disease (atherosclerosis), inflammatory bowel disease, psoriasis, autoimmune thyroiditis, Alzheimer's disease, schizophrenia, myalgic encephalomyelitis, rheumatoid arthritis, lupus, COPD, celiac disease, and obesity—among a much longer list. It is no wonder that prescriptions of medications for depression are at an all-time high.

To date, treatment of NCDs has focused more on managing symptoms of single diseases rather than correcting the underlying biological cause that connects multiple diseases in a pattern. This strategy is like a lazy approach to home repair. Suppose that during a major storm, your house develops a leaky roof. Water from the heavy rain seeps into the interior of your house and damages a portion of the ceiling in one room. You repair that spot in the ceiling as best as possible, being very pleased with your effort. The next rainy season, you notice that ceilings in two rooms have water damage, as does the Sheetrock in a section of wall. Ever diligent, you repair both ceilings and the section of wall. The third year, the leaks are so large that the floors are getting wet. You make all the ceiling, wall, and floor repairs as before, replacing the carpets and hoping against hope that mold does not become a future problem. Is the leaky roof likely to get less leaky?

If the root cause linking multiple NCDs together is never addressed, the core problem producing the diseases remains in place, and there are likely to be additional NCDs in your future. That is one reason why we have this epidemic. At the center of it all is the microbiome, because it is the driver of immune maturation and the capacity to determine whether we have healthy or unhealthy inflammation.

The microbiome is like the roof and exterior walls of your house. It is the go-between between you and your exterior world.

4. Microbiome Status Affects NCD Status

Your microbiome status can affect both if and when an NCD will develop and whether current treatments are likely to be effective. Evidence shows that a dysfunctional microbiome (called microbiome dysbiosis) is predictive of certain NCDs. In fact, different NCDs can have their own specific microbiome profile, a type of fingerprint associated with them. But does dysfunction within the microbiome directly cause NCDs, or does it merely help to lock them into your physiology, making corrective therapies very difficult? We don't yet know for sure, but to find out we will have to experiment and include the microbiome in therapeutic approaches.

Right now researchers are specifically investigating: (1) Which microbial imbalances cause or lock in which diseases? And (2) how can a compromised microbiome in one part of the body (gut, for instance) lock virtually every NCD into place in unrelated locations throughout the body (brain, for example)?

Below I list thirty-two NCDs. This is just a partial list of the known NCD-microbiome connections. But the combined breadth, range, and impact of these diseases show the importance of working through the microbiome to attack the NCD epidemic. For each NCD, research has established that the disease is tightly associated with a dysfunctional or incomplete microbiome. In some cases, the problematic microbiome seems to show up first and the disease later.

Here they are:

Alzheimer's disease
Asthma
Autism
Autoimmune hepatitis
Breast cancer

Cardiovascular disease

Celiac disease

Chronic kidney disease

Chronic obstructive pulmonary disease (COPD)

Colon cancer

Crohn's disease

Depression

Food allergies

Hypertension

Laryngeal cancer

Lung cancer

Lupus

Nonalcoholic fatty liver disease

Obesity

Osteoporosis

Parkinson's disease

Periodontal disease

Prostate cancer

Psoriasis

Respiratory allergies

Rheumatoid arthritis

Schizophrenia

Sudden infant death syndrome (SIDS)

Type 1 diabetes

Type 2 diabetes

Ulcerative colitis

Urothelial cancer

For many NCDs, transferring the microbes or their metabolites from a diseased individual or animal to one that is healthy can reproduce the disease in a recipient. For example, in rats, obesity can be fully transferred with just the gut microbes. The transfer causes the unsuspecting recipient to become obese. Each NCD requires drug therapy to manage symptoms, often for life. In the case of obesity or diabetes, anthocyanins from black currant berries can help normalize glucose metabolism and produce weight loss. But in studies in mice, these antiobesity chemicals in berries only work when a complete microbiome is present. Interestingly, a recent study from Korea shows that the Chinese weight-loss drug ephedra actually works by altering the microbiome. These results suggest that trying to eat a healthier diet to lose weight or reduce the likelihood of diabetes will work best after you first install a healthy microbiome. Leaving the microbiome defective may make dietary interventions ineffective.

The take-home message from research on these thirty-two NCDs is to go after the microbiome first. Installing a healthy, intact microbiome makes it much more likely you can protect against and reverse NCDs. Otherwise, a dysfunctional microbiome will cause you to metabolize your food in a less than useful way and will always be pushing the immune system toward NCD-promoting haphazard immune and inflammatory responses in the tissues.

Statins are the most prescribed drug to treat high cholesterol. High cholesterol can be a problem because it often leads to cardiovascular disease. We know that microbiome status affects the risk of cardiovascular disease. Lo and behold, it also determines whether or not statins will work for each individual. Gut microbes metabolize statins, altering them before they ever reach the body's tissues. When the microbiome is deficient or imbalanced, statin metabolism can change so that too little of the actual drug reaches the blood to be effective in treating the condition. Something as simple as a course of antibiotics given for an infection can sufficiently alter the microbiome to wreak havoc on NCD drug therapy.

Digoxin, a drug derived from the foxglove plant and used for centuries to treat heart disease, is another example. Because digoxin can be toxic and lead to death, meticulous dosing is required. Too little digoxin and it has no effect on heart function; too much digoxin and the patient dies. A critical gut microbe for digoxin metabolism is *Eggerthella lenta*. Without just the right numbers of bacteria, digoxin metabolism is compromised. In a patient with too many of these bacteria, the drug is completely inactivated in the gut. With too few bacteria present, the prescribed dose of digoxin could be a lethal overdose.

Even more so than with other conditions, the microbiome rules when it comes to cancer therapies. If your microbiome is not intact, three major classifications of cancer drugs absolutely cannot work in the body. These drugs include oligonucleotide therapies, platinum chemotherapy, and cyclophosphamide chemotherapy. Yet it is doubtful that any oncologists check the microbiome status of their patients before beginning cancer treatment. That is unlikely to continue.

To summarize, NCDs are all interconnected through uncontrolled inflammation and risk of comorbid diseases. The microbiome is at the heart of our NCD epidemic. A compromised microbiome programs immune dysfunction and misregulated inflammation, just like you program your DVR for recording a certain TV show or movie to watch later. It also locks the NCD into your basic physiology, making the disease more resistant to dietary and drug therapies. Finally, microbiome dysbiosis helps promote more disease. It paves the way for additional NCDs, more medical dependency, and reduced quality of life. And once an NCD appears, the microbiome needs to be complete and balanced to ensure effective drug therapy.

9

THE SIX CAUSES OF THE EPIDEMIC

NCDs run rampant across the globe, and we only now seem to be getting a clue about fundamental aspects of human biology—was there some awful conspiracy? I don't think so. There were no colossal errors or horrific judgments that got us here. Hindsight is always easy. Instead, there were well-intentioned people. In almost every case, people were drawing upon new inventions, new ideas about what could be helpful for humans, new opportunities for jobs and affordable housing, and they were helping societies make progress as well as they knew how. The rewards for decisions that were made primarily during the twentieth century were immediately obvious. The risks were never realized until the twenty-first century.

The combined changes in our diet, living conditions, medical treatments, and approaches to human safety had downsides that are only now becoming apparent. Professionals were either not aware of or greatly underestimated the risks associated with various practices and lifestyles. Health guidance that was provided by professionals failed to recognize that such practices could damage the microbiome and, in turn, make our immune systems a living nightmare of self-destruction, cancer, and disease. These professionals simply did not know. It is sad but understandable given the old biology.

Now, sitting in the middle of the NCD epidemic, we can easily look back and understand how seemingly useful and harmless practices ushered in disease and disability. Many contributing factors led us down the path toward more and more people growing up with a deficient or damaged microbiome and later-life NCDs. There are six prime factors that led the way to this pivotal point in health and medicine:

1. Antibiotic overreach
2. The food revolution and diet
3. Urbanization
4. Birth delivery mode
5. Misdirected efforts at human safety
6. Mammalian-only human medicine

Unfortunately, to date, these practices have yet to be overturned or adequately modified. In some cases the risks are better known now. But only baby steps have been taken to address them. Making changes at both personal and institutional levels will be a critical part of any new medicine.

These are not the only factors involved. However, they are the factors we will need to address in the medical revolution if we are to right the ship of human superorganism wholeness and health.

Here is what we need to do.

1. Antibiotic Overreach

Antibiotic overreach includes both the inappropriate use of antibiotics (as a routine supplement in animal feeds and for probable viral infection in humans) and an underappreciation of the costs to health involved in antibiotic use. While antibiotics saved lives during the twentieth century and continue to today, too much of a good thing is often a problem, and that is certainly the case with antibiotics. But this problem is not like a child getting a tummy ache from eating too much ice cream. No, this is a much bigger, longer-lasting problem. Inappropriate antibiotic use is more like taking an action that results in losing one of your organs or an arm or a leg. The cost to your health is high, so it had better be worth it.

In a recent study, the antibiotic amoxicillin was found to be the most frequently prescribed drug for US infants and children. How did that even happen? It happened because if your child has a bacterial infection, antibiotics can usually clear it. They work. Physicians assumed that the worst that could happen by prescribing an antibiotic in the event the infection was caused by a virus was that it would not work and might contribute a bit to antibiotic resistance among bacteria. But the thinking was that inappropriately prescribed antibiotics had no real downside for the patient. No harm, no foul was the idea. But we now know there is harm and potentially a significant amount of it. Under the old biology, bacteria were generally seen as evil, and widespread killing of our bacteria by antibiotics was no problem. One round of antibiotics can damage your microbiome and cause your entire metabolism to change, along with the interconnected functions of your tissues and organs. How could people know that killing off more of your bacteria than just the one causing the illness was actually destroying an important part of you?

Originally, antibiotics were used when illness was a matter of life or death such as with cholera, typhoid fever, staph infections from wounds, or tuberculosis. But now they are used far more routinely in doctor's visits such as with infant ear infections, called acute otitis media (AOM). This is despite the fact that a high percentage of those infections are caused by viruses, which are not susceptible to antibiotics. The majority of these prescriptions are administered by general practitioners. Such overuse has resulted in pressure to discontinue this practice when possible, in part because spontaneous remission of AOM without complications often occurs.

Overuse of antibiotics also increased antibiotic resistance in pathogens, giving rise to methicillin-resistant *Staphylococcus aureus* (MRSA) infections and other so-called superbugs. The threat is real and growing. The US Centers for Disease Control and Prevention recently estimated that at least two million people in the United States become infected with superbugs annually, resulting in tens of thousands of deaths each year. As a result, there is now a race to develop new antibiotics that will work. Yet all of this is happening without an effective medical plan to replace the nontargeted, beneficial bacteria that are also destroyed during antibiotic treatment. We did not want to lose these beneficial bacteria nor can we really afford to lose them. Antibiotic treatment of children can have several major effects: loss of key bacterial species that are required for the entire microbiome to mature correctly, loss of diversity within the microbiome, loss of key metabolic functions needed by mammalian cells and tissues, and increased later-life susceptibility to serious infections. Losing useful bacteria opens up space for harmful pathogens to move in and claim our body's territory as their own.

The health problems linked to children who have lost major parts of their microbiome are things like obesity, diabetes, cardiovascular disease, neurological problems, allergic and autoimmune diseases, depression, and cancer. In the past, we didn't know about the health ramifications of damaging our microbiome. But we do now.

The problems stemming from antibiotic overuse were not restricted just to humans and prescription drugs. My first responsibility as a newly minted Cornell professor was to develop naturally healthier chickens. I got firsthand experience and insight into the world of globalized food production and sustainable agriculture. Poultry represents the number one meat protein source consumed not just in the United States but globally as well. To paraphrase a famous Cornell professor and public educator, Carl Sagan, there are billions and billions of them, but in the case of my early career, they were chickens rather than stars.

One thing I quickly learned is that, in both the research strains and agricultural production chickens, we control virtually every aspect of their environment. We determine what they eat, drink, and breathe (i.e., air quality), as well as their housing type, space, vaccinations, and lighting. This exquisite control and the desire to gain optimal production (whether in terms of eggs or meat) led to the discovery in poultry of many of the essential vitamins and minerals our bodies need. In fact, Leo Norris, a foundational nutritionist from Cornell who worked on poultry, first described riboflavin and magnesium deficiencies from observing their effects in chickens. My observations and thoughts about Cornell poultry were generally far more pedestrian. I once worked with a veterinarian who had previously treated the Triple Crown–winning racehorse Secretariat, and wondered why our specially bred birds did not have the same monetary value as a pedigree thoroughbred racehorse or at least a prize bull.

At the end of World War II, there was a significant need for cheap animal protein sources. It was the same decade when poultry feeds were being supplemented with both folic acid and vitamin B12. In this mid-twentieth-century climate of animal feed supplementation, a decision to include antibiotics in poultry and livestock feeds was viewed as simply including just one more additive that would enhance growth and/or productivity. Folic acid, B12, and antibiotics were lumped into the same general category as modifying the diets of production animals. Now, almost seventy years later, we can see what damage having only part of the scientific story (the old biology) can do.

This became the widespread scourge known as routine use of antibiotics in animal feed, which were everywhere and in every feed. The antibiotic drugs took a variety of chemical forms, including some with toxic substances such as arsenic. Even when the levels of antibiotics used in feeds were subtherapeutic (using lower doses than would be used for treating actual bacterial diseases), they were still being released both into the animals and into our environment, causing pernicious effects. By the 1950s, studies reported bacterial resistance against the antibiotics used in poultry feed. The genes that conveyed this resistance did not stay just in poultry flocks but could be transmitted to microbes that could infect humans.

Alarm bells began to sound as early as the late 1960s in the UK. But it was not until the 1980s and later that US science and health organizations began taking note of the potential risk. Sweden was the first country to ban antimicrobials for the purpose of promoting animal growth (1986), and a series of antibiotics were banned for that purpose by Denmark during the 1990s. After 2000 the World Health Organization initiated global opposition to using antibiotics in animal feed. But there was little regulatory activity in the United States. Some producers began to reduce or eliminate the use of antibiotics voluntarily in the face of the accumulating scientific evidence pointing to a larger environmental health concern. But it wasn't until September 2014 that one of the largest US producers, Perdue Farms, announced it would no longer use antibiotics for growth promotion via egg injections. However, that still leaves the issues of antibiotics in feed, and some reports suggest that antibiotic use in feed continues among major producers.

More than twenty years ago, near the end of my poultry career, I became concerned about various practices in animal agriculture, including the routine use of antibiotics in animal feed. I coauthored a paper advocating for a dietary and natural immune approach to animal agricultural management that would reduce bacterial pathogen load during production, and followed that paper up with arguments against antibiotic use in animal feed in a June 25, 1998, article published in the *Christian Science Monitor* and other media sources. At the time, I was among an increasing number of scientists speaking out against the routine use of antibiotics as an agricultural food supplement. The three major points I made nearly twenty years ago were:

1. Antibiotic resistance is real, and the biology of the process alone told us we should not put antibiotics from billions and

billions of chickens into our environment and food chain when the growth and health of chickens could be achieved in other less damaging ways.

2. Massive antibiotic administration via animal feed was an unnatural defense process for the chickens themselves. The animals were not being bred and managed based on an integrated natural health plan, but were being loaded up with drugs for as long as was allowed by regulations. Because the chickens were constantly being fed antibiotics, their immune systems never got exposed to infectious bacteria and never developed a protective immune response against them. Part of the normal immune response against pathogens caused muscle loss. Producers hate muscle loss because muscle loss means less meat per bird and lower profits. Antibiotics were given in feed in part to produce chickens with larger breasts. But this strategy meant that far less attention would go to actually having healthy chickens across all breeds (laying birds and meat birds) and stages of life. If you could just add more and more chemicals and drugs to the birds and not worry about whether they had innate resistance to infections and other diseases, that became a lower priority in both breeding and environmental management. A "we'll just drug them up" approach took over, giving rise to an ever-increasing reliance on more and more chemicals and drugs that would enter our food chain.

3. The massive antibiotic approach was incomplete. Regulations require that the drugs be removed from the animal days to weeks before they are processed for your dinner table. This was necessary to reduce the levels of antibiotics remaining in the meat and/or eggs to what were deemed safe levels for the consumer. Of course "safe" was determined before we understood what constant low-level antibiotics coming through the food

chain might do to our microbiome. Because the poultry producers had been all-in for antibiotics, once they were removed, the
animals (and the farmers' livelihoods) were vulnerable to infections and diseases. Since inherent immunity of the animals had
not been needed for the vast majority of the animals' lives, there
was a greater likelihood that the animals at the processing
plants would be chock-full of newly emerged infectious agents
that took over once the antibiotics were removed. A race was on
to see if the animals could just make it through the processing
plant carrying newly emerging infections before the spread of
infection required a high percentage of birds to be condemned.
But the newly growing bacteria would still be there since the
antibiotics-for-all approach did not stress the value of the bird
having natural resistance to disease. An increasing number of
those bacteria present at processing would be antibiotic resistant as well. Antibiotic residues for your own microbiome and
aggressive bacteria interested in your gut were probably not at
the top of your list when you last shopped at the grocery store.

In the late 1990s I noticed a trend among the more visionary and influential poultry farmers in New York State. They voluntarily chose to wean
their farms off of antibiotics in feed. Farm by farm, there has been a grad-
ual shift under way to a different type of environmental management of
poultry microbiomes. Antibiotics in animal feed have been on their way
out, even if consumers had to make it so with their purchasing choices.
Regulatory agencies largely stood on the sidelines watching.

2. The Food Revolution and Diet

In general the world has never had so much food and yet has never
starved our microbial partners so much. It is an unusual story of grab-

bing failure from success and how through technology we have created a wonderland of food but have gotten it very wrong when it comes to what we really need to eat as a superorganism.

I grew up between the early 1950s and the '70s, during a perfect storm for food in the United States. It was a time of amazing technological advancement and the shedding of many old country traditions that were rooted in local crop consumption and the need to store food for survival through harsh winters. It was a time when my family went from pre-TV to having a black-and-white TV, and during a shift from few frozen foods to many frozen foods as fully prepared frozen dinners supported two-career parents. (We won't talk about the composition and taste of some of those early, complete frozen dinners.)

The interstate highway system project under President Dwight Eisen-hower had led to new opportunities for moving people and, most impor-tant, food. Early attempts to transport food without it spoiling relied on dry ice. But a breakthrough in 1939 changed how we access and connect with our food. In that year Cincinnati-born Frederick Jones filed a major patent. Jones was a largely self-taught inventor and the holder of sixty-one patents. Most of his inventions involved sound equipment for mo-tion pictures or innovations in refrigeration. It was Jones, along with his partner Joseph Numero, who developed a mechanical system called the Thermo King for refrigerating tractor-trailer trucks. Most of the recog-nition for Jones's remarkable contributions was received posthumously. In 1991, he became the first African American to be awarded the US Na-tional Medal of Technology. The Thermo King refrigeration units were attached to the underside of the truck trailer. With the improving high-way system, semis could begin to roam the country, delivering meats, fruits, dairy products, and vegetables with shorter delivery times. For the first time, food grown elsewhere could be stocked on shelves every-where.

The first half of the twentieth century involved train transportation of food by railcars that were iced. Many boxcars were really iceboxes on rails. This was not particularly efficient and required teams of icers at icing stations along each rail route. The icers were mobilized much like today's teams of pit crews for auto races. They would refill bunkers on each car with ice. These bunkers were usually accessed through hatches on the roof. The system was labor intensive and still had its limitations. By the mid-twentieth century, mechanically refrigerated train cars, essentially Thermo King units for trains, were an important technological development that gradually replaced ice-packed transportation. This paralleled what was happening in the trucking industry, not just in the United States but also in other countries.

Thermo King self-contained refrigeration units could be moved from trains to trucks and even to ships, opening up marine transport as well. Food and other perishables, including medicines, could be moved vast distances under refrigeration. In the 1950s, there was less pressure to use fermentation to store food safely, and the fermented foods, so important to the health of our ancestors, lost their place on our dinner tables. The technology was wonderful, but we had no clue that we were literally losing a microbial part of ourselves in the process.

Frozen foods became another convenient food choice, another option when the range of fresh foods was limited. That technology was aided by New Yorker Clarence Birdseye's invention and commercialization of the flash-freeze process. He had watched indigenous Arctic people such as the Inuit use quick-freeze methods for preserving fish and realized that he could simulate those conditions, preserving the freshness, form, and palatability of food when thawed months later. The technology eventually found its way to a later-formed corporation, General Foods.

Because both refrigerated and frozen food could now be shipped across the United States, in my 1950s childhood San Antonio home we could enjoy apples from New York and Washington state, potatoes from Idaho, peaches from Georgia, avocados from Texas's Rio Grande valley, and berries from California. But there was an absence of probiotic-containing food everywhere one looked. Of course we weren't really looking either. Like most in the 1950s in the US, we were enjoying the food revolution, the increased availability of food choices across seasons, without realizing that a fundamental food component was being lost: probiotic microbes. In fact, with yogurts and kefir not yet common in the US, it was a biological wasteland for the human superorganism. Never before in human culture had our diet options been so extensive yet contributed so little to our internal biological diversity.

Of course the intention of the inventions providing this ready access to even nonseasonal food and a reduced need for long-term storage methods of the past was that we could avoid starvation, work longer hours away from our homes, and keep ourselves relatively healthy. But what we did not realize is that the food of our ancestors, much of it stored using fermentation to protect against agriculturally barren winters, was better at supporting our microbiome. It was feeding us not only nutrients but also what we now know as probiotics. Additionally, many of the foods contained what are now called prebiotics, food components that feed the majority microbial part of us. In Part Three of this book, I will discuss prebiotics and probiotics in more depth.

Because we were unaware of the importance of our microbiome, including its care and feeding, a paradox was established. Food was now available year-round, and food could be shipped from areas of plenty to areas facing starvation due to drought or conflict. But in the changes we were making, we were unknowingly depleting and starving our commensal microbes. We unintentionally became incomplete as a superorganism and increasingly dysfunctional and unhealthy as a result.

Food choices are interconnected with the food and agricultural revolutions that have changed much of how our food is produced, stored, and made available. In addition to what we are no longer eating, we are eating some things our ancestors either did not have available or chose not to eat. That is not to imply that the food of our ancestors was always inherently better. It is simply to say that in one or two generations we changed our diet in unprecedented ways. Some of our food choices are also linked with the last category of this chapter: safety. In practice, food safety has usually meant the elimination of something that could poison the individual or cause infections (e.g., pathogenic bacteria). This is a place where impact on the microbiome has not been considered until the past few years, and there is a lot of catching up to do.

For people who are often away from home, convenience is a big part of food availability. A great deal of research has gone into food processing to create tasty products almost ready for consumption rather than lots of raw ingredients requiring hours of preparation. In hindsight, the processing may have come at a cost when it comes to complex formulations that can have unanticipated or undesirable effects on the consumer. If the full extent of breast milk's positive effect on the microbiome was misunderstood, then it is safe to say that processed foods are a big unknown relative to their impact on the microbiome. In cases where they have been examined, the findings raise concerns.

When it comes to promoting the health of the superorganism via diet, there are several recent research-based findings. First, many of the vitamins we need are produced by the lactic acid bacteria in probiotic foods as well as by specific commensals in our gut such as bifidobacteria. Because our mammalian cells do not make most of our vitamins, at a minimum we should seed our microbiome and then eat a diet that feeds the gut microbes that produce the vitamins we need. Stanford researchers recently described the lack in many diets of critical carbohydrates that are needed by our microbiota. In fact, they argue that we have essentially starved out our useful microbes with a westernized diet.

That is certainly a part of how we got to where we are now. But the good news is that dietary components supporting needed gut microbiota are now well known, as are foods that are harmful for our microbiome. Truly holistic diets that support the completed self can be pursued.

3. Urbanization

Urbanization has had an interesting and varying effect on human health for centuries. An expansion in urbanization increases health risks, although the reasons for this happening today are somewhat different than in the past. In the past, unsanitary conditions, combined with people huddled together, allowing easy spread of infectious diseases, made cities a perilous place for health. Today, many of the aspects of modern city life are the apparent causes of a disrupted microbiome; a dysfunctional inflammation-prone immune system; and a totally different set of diseases.

The trend to move to cities is nothing new. One of my scholarly hobbies is research on Scottish history, in particular the history of the goldsmiths of Edinburgh, Scotland, where the Incorporation of Goldsmiths has existed for more than five hundred years. You can learn a lot about urban living from the records and stories of highly skilled craftsmen who lived and worked in a town's center. Edinburgh is an ancient city that was originally built on steep hillsides above a swamp, the whole of which was defended by its monumental castle from the early 1100s. Scotland itself had a largely agrarian culture until the Industrial Revolution. Extant historic details of city life as well as city death can be quite revealing.

During the seventeenth and eighteenth centuries, Edinburgh was a busy town with overcrowded tenements and various places of business crammed into tall buildings along a few streets. These tall buildings were even placed along the side of the largest church, St. Giles' Cathedral, sitting in Parliament Square along the Royal Mile running from the castle to Holyrood Palace. In its heyday the Royal Mile of Edinburgh was like a mini Manhattan, only with more wood. People of various social classes were intermingled, and residences and shops were all nearby in the cramped spaces. The town's affectionate nickname was Auld Reekie (probably meaning Old Smoky due to numerous wood and coal fires but often interpreted to have meant Old Smelly). It was a very polluted and unsanitary hill. Human waste was thrown out of windows of dwellings onto the streets, eventually flowing down the streets to the swamp. You wanted to live as high up in those buildings as possible. Infectious diseases were rampant.

Among these filthy conditions there were many things of beauty being crafted. One of the eighteenth century's most famous Scottish goldsmiths was the second-generation master James Ker, who was noted for his marvelous gold and silver wares. James Ker was famous for navigating the perilous political waters following Scotland's 1745–46 Jacobite rebellion ending with the Battle of Culloden. He rose to the lofty status of representing Edinburgh in the British Parliament in 1747. In those days, election to Parliament was a position normally reserved for the town's most well-respected merchants.

James Ker's family life almost never happened. His goldsmith father, Thomas Ker, lived in a cellar under his shop in Edinburgh's Parliament Square, with the only outside ventilation coming through a grate that opened directly into the sewage-draining side of the steep street. It was described as a "miserable and unhealthy hovel." Of thirteen children, James was the only son to survive infancy—and only because the family had relocated to healthier housing. Ironically, James Ker had little more success raising the next generation of Ker children to adulthood, though he and his wives certainly tried. He had twenty children between two wives, with only five known to have survived. The majority of those surviving childhood were born after James Ker was wealthy enough to buy a country estate called Bughtrig. His daughter Violet married his talented apprentice William Dempster and forged the basis of a long, lucrative partnership while Ker was in politics and in London. One of his few surviving sons, Robert Ker, became a famous science writer. Robert was born away from Edinburgh on a country estate where his mother had been at the time. That quite possibly saved his life.

Of course, in Edinburgh during the two generations of Ker goldsmiths

(1650s–1760s), infectious diseases (communicable diseases) were the leading cause of death. In stark contrast, in 2013 the leading causes of death in Edinburgh were noncommunicable diseases like cancer and circulatory diseases, which made up more than half of all deaths.

What happened in Edinburgh between the eighteenth and twenty-first centuries is a microcosm of what is happening all over the world, and overcrowding is still involved. City planning took over in nineteenth-century Edinburgh with the draining of the swamp and construction of New Town to accommodate more people wanting to urbanize. Sanitation improved with these changes. As a result, instead of massive amounts of human waste causing widespread infections and death, it is the chemical by-products of urban human activities that are now assaulting and breaking down our bodies.

More people than ever live in urban areas and major cities compared with rural areas. According to a 2014 United Nations report titled "World Urbanization Prospects," 54 percent of the world's population now lives in urban areas, and this is expected to increase to 66 percent by the year 2050 in association with the creation of new megacities, each with more than ten million people. Cities have been a massive drawing card for jobs, services, and a variety of entertainment. The general formula has been that more people per square mile equals more jobs, more commerce, more transportation opportunities (if you need convincing, just compare the destinations available flying out of metropolitan New York City airports with those of regional upstate New York airports), and more stuff to do. The centuries-old demand for urban living has led to modern-day university degree programs and jobs for urban planners to create integrated city spaces to accommodate all of the activities of large populations of people. These human-made spaces have been called the "built environment." Considering the amount of planning that has gone into cities like San Francisco, New York City, Tokyo, Beijing, Seattle, London, Rome, São Paulo, one would think they must be the healthiest places on earth. Well, not exactly.

A major concern is that urban planners may have all the bells and whistles you could think of surrounding a major metropolis, such as services, entertainment districts, green spaces, hiking and cycling trails, parks, residential-density planning, public transportation, etc., but they have somehow missed a key asset for those choosing to live in the city: protection against the world's most common killer, NCDs.

Recently, researchers have recognized that people living in urban areas, regardless of the city or continent, have significantly elevated percentages of noncommunicable diseases compared with those living outside cities. This can include age groups that normally don't see high rates of mortality. Professor Arline Geronimus of the University of Michigan Population Studies Center noted that young to middle-aged residents of some impoverished areas within cities suffer extraordinary rates of excessive mortality where chronic disease contributed heavily to the deaths. Of course the question is why? There are some pretty good leads.

Several components associated with urban living could contribute to problems with the microbiome as well as increased risks of noncommunicable diseases. One of the most studied is air pollution. The fine particulate matter (PM) concentrated in urban air is a significant concern. Researchers have associated exposure to PM with elevated systemic inflammation. As described in the prior chapter, inability to shut down inflammation when appropriate is a major component contributing to onset or maintenance of NCDs. Specifically, living near major roadways, as happens in all major cities, significantly elevates the risk of both heart disease and asthma. For asthma, the developmental timing of the exposure and the sex of the child influence the risk following exposure to the air pollution of cities. Additionally, there is evidence that exposure to traffic-related urban air pollution increases the risk of obesity. Again the promotion of inflammation by urban air pollution is thought to be involved.

If normal metropolitan areas are not enough of a concern, China is moving to create megacities with unprecedented numbers of people and related activities concentrated into comparatively small geographical areas. Two megacities are planned. One in the Pearl River delta is intended to have more people than Canada or Australia; another involving Beijing, which will be called Jing-Jin-Ji, will have an estimated 130 million people. Is it a good idea? The answer might depend upon what one wants out of it. Anyone who witnessed the air pollution connected with the recent Summer Olympics held in China can imagine what a multifold increase in the concentration of particulate matter would mean for human health.

Rather than creating megacities that would be expected to increase destruction of the human microbiome and elevate rates of NCDs even further, a return to a lower-density, country-type life might be a more healthy direction. That fits with a scientific idea that has been discussed for some time in various forms, called the hygiene hypothesis. In fact, the urban versus rural effect has been known for some time in terms of impact on the immune system and risk of multiple NCDs. It might have you singing, "Farm livin' is the life for me" like Eddie Albert in the 1960s show *Green Acres*.

Green Acres was the poster child for country living in the small town of Hooterville, where Eddie Albert, Eva Gabor, and of course Arnold the Pig lived. (If you are too young to remember that show, just wait: Both a Broadway musical and a movie version are planned.) Beyond air pollution, the urban environment removed us from contact with farm animals, the environment, and the food that supports both the microbiome and a well-regulated immune system. The urban versus rural effect was first noticed in comparisons made in Germany on the health risks for children who grew up on a farm compared with those in a nearby city. Despite other factors being similar, the farm versus city living was associated with significant differences in risk for a specific category of immune-related NCDs: allergic disease and asthma. At the same time, the farm versus city environment has been shown to affect immune development. In a recent article in the journal *Science*, researchers from Belgium and the Netherlands showed at least part of the basis for the immune-protective effects of early life on a farm. They found that the microbial products in dust from farms with animals can program the immune system differently, such that it is better balanced, generates less unhealthy inflammation, and requires a higher threshold of allergen exposure before any type of allergic response would be produced.

Urbanization is a superorganism health problem. That can be changed, but it means radically altering how cities are structured and operated. Somehow we must either change the environments of cities or move out to the country.

4. Birth Delivery Mode

The method by which a baby is born is one of the most significant factors affecting a baby's microbiome. While newborns are exposed to some bacterial products during prenatal development, the single most important seeding event for the microbiome is at birth through vaginal delivery. That provides the foundation for microbes throughout the body (mouth, gut, urogenital), and they will co-mature with the developing immune system. Not surprisingly, birth delivery mode is also a significant factor affecting the newborn's immune status and risk of noncommunicable diseases.

When a child is delivered by cesarean section, the microbiome is not properly seeded and adequate microbial colonization is delayed unless a complementary therapy is used. With our understanding of the new biology and the view that self-incompletion is essentially a birth defect, the consequences of not establishing our majority microbial cells and genes at birth are becoming obvious. Cesarean delivery can be medically necessary and should be used when that is the case. However, in a recent journal article I had the opportunity to consider the origins of cesarean delivery and the evolution of the practice across the years, leading to the historically high rate of global cesarean deliveries in the present.

Cesarean delivery was originally employed only to save the baby when the mother had just died or was dying. In fact, an ancient Roman decree called the *Lex Caesarea* stated that cesarean delivery would be attempted on all dead or dying women who were with child. The idea that it could be used with both mother and baby surviving is a comparatively modern use of the medical procedure. The first purported instance of mother and baby both surviving came from Switzerland in 1500, although the event was not recorded until the 1580s.

Once antiseptics, anaesthesia, and antibiotics made cesarean delivery survivable and safer, its use as the birth mode of choice exploded. This opened up the possibility of elective operations. Planned delivery dates had certain inherent advantages for all parties involved. With everything paving the way for elective cesarean deliveries as an option, if not preference, there has been a steady rise in elective cesarean deliveries in both developed and developing countries. There was a 53 percent increase in cesarean deliveries in the United States between 1996 and 2007. The increase occurred across all states and ethnic groups. In Sweden there was a threefold increase in elective cesarean births just between the years 1997 and 2006, and in England the rate doubled between 1990 and 2008. Recent reported rates for cesarean delivery are as follows: England, 24 percent; US, 33 percent; parts of India, 40 percent; Brazil, 32 to 48 percent, depending on the mother's country of origin; and China, 46 percent. Of course, the massive rise in elective cesarean deliveries was all based on the assumption that there was little downside to the procedure, particularly once immediate surgical risk was past. But the risk-benefit estimates were wrong because the understanding of biology and approach to safety testing were wrong. Now we know better.

Invariably, throughout the twentieth century we have used short-term measures in determining what is safe. That is fine when we are considering infectious diseases, pandemics, and acute poisonings. But it is far from adequate when considering lifelong safety. It is almost as if we have been willing to check a week after an operation or environmental exposure, and if no problem could be detected, then the event was considered to be safe. But if the new biology teaches us anything, it is that what you measure and when you measure are both crucial. Merely waking up the morning after a dinner hosted by the ruthless Lucrezia Borgia might come as a relief, but that measurement of survival is not the best predictor of health across a lifetime (or even a month in the dangerous world of Italian Renaissance politics). Developmental programming happens, epigenetic regulation happens, and no-see-um issues with the microbiome, the immune system, and the neurological system happen much like a ticking time bomb. We would never know there was a problem just by looking at the newborn given the usual measures that have been used in Western medicine and safety evaluation.

The missed health risks connected to cesarean delivery are twofold, based on the latest results. First, the surgical procedure, like most, includes presurgical administration of antibiotics to prevent postsurgical infections. The antibiotics compromise, if not destroy, the mother's microbiome, which needs to be passed to the baby, and also impair bacterial signals the baby is receiving from those maternal microbes just before delivery. Essentially, a mother is given a drug affecting 99 percent of the genetic component she should be passing on to her baby, and in standard medical practices, to date, nothing is done to correct this. Then the C-section bypasses the transfer of the coating of microbes from the mother's vagina to the baby so they can seed the baby's gut. Cesarean delivery interferes with the birth of the human superorganism. This presents future health challenges if nothing is done to biologically complete the baby.

The timing and nature of the founding microbes in body locations like our gut are critical for both our physiological maturation and the developmental programming of later health. Fredrik Bäckhed and research colleagues of Sweden recently compared mother and infant microbiomes across the first year of life. Their conclusions? Below are the highlights.

1. A baby born vaginally has a microbiome that looks like its mother's. Based on analyses of stool samples, bacterial species in the baby's gut were a 72 percent match for those in the mother's gut, while the bacteria in cesarean-delivered babies had only a 41 percent match with those in their mothers.

2. More bacteria in cesarean-delivered babies were derived from sources outside the mother (e.g., hospital workers and surfaces) as well as from the mother's skin and mouth. However, the bacteria in the skin and mouth are not the ones normally needed in the lower gut to promote co-maturation and the most effective metabolism. The profile in vaginally delivered babies features genera *Bacteroides*, *Bifidobacterium*, *Parabacteroides*, *Escherichia*, and *Shigella* bacteria. In contrast, C-section-delivered babies were installed with genera *Enterobacter*, *Haemophilus*, *Staphylococcus*, and *Streptococcus*, and *Veillonella* bacteria.

3. As the cesarean-delivered babies developed, they were missing or had much fewer *Bacteroides* bacteria compared with vaginally delivered babies.

4. As the babies aged, the microbiome of cesarean-delivered babies appeared to look more like that of adults sooner than the vaginally delivered babies' microbiome. It is as if with the C-section they had missed some earlier developmental progressions. At each developmental stage during the first year of life, the C-section signature bacterial species differed from those in vaginally delivered babies.

5. The microbes in cesarean-delivered babies carried a greater proportion of genes for antibiotic resistance than those of vag-

inally delivered babies when first born, and the difference was still significant at four months of age. That could affect the babies' capacity to receive effective future treatment with antibiotics. In many ways, that is not surprising since more of the microbes in cesarean-delivered babies either came from or had more exposure to the hospital environment than those in the mother's birth canal.

6. The early infant microbiome, when complete, is designed to receive and process breast milk as the initial food source.

7. Metabolism by the early infant microbiome is central to the production of key vitamins, iron, and amino acids required by the brain for its development.

An editorial article accompanying the Bäckhed group's paper on Swedish moms and their babies was titled "Birth of the Infant Gut Microbiome: Moms Deliver Twice!" I absolutely agree with this.

An additional, recently described effect of cesarean delivery involved an attempt to use maternal probiotics to boost microbial transfer to the baby through both colostrum and breast milk. A collaborative research group from Italy examined the effects of giving a daily probiotic mixture containing lactobacilli and bifidobacteria to women during late pregnancy and early lactation on the microbes found in both colostrum and breast milk. The probiotic bacteria levels were significantly elevated in both colostrum and mature breast milk in the women who had delivered vaginally, but there were no significant increases in the probiotic bacteria in the colostrum or breast milk of women who had given birth by C-section. In this case, birth delivery mode influenced the levels of probiotic-ingested bacteria that were subsequently available for transfer to the baby through colostrum and breast milk. That was an unexpected finding.

There certainly are associated effects of cesarean delivery on both immune maturation and risk of many NCDs. Also, the types of problems with the immune system suggest they underlie the elevated risk of NCDs. For example, one of the needed changes for the newborn's immune system is for the Th1 branch of acquired immunity to catch up with those types of responses promoting allergic diseases (Th2). There is a bias toward Th2 prenatally, and that imbalance has to be corrected through further maturation after birth to provide infants with immune balance. In general, Th1 responses are most useful in fighting viruses and cancer while Th2 responses are the biggest help fighting parasites and certain kinds of bacteria. Ultimately, the infant needs both types of responses in balance to fight diseases and maintain tissue integrity. Imbalances between Th1 and Th2 capabilities usually end in disease.

Researchers at the Swedish Institute for Communicable Disease Control showed that cesarean delivery not only causes problems with the gut microbes, but also keeps the Th1 branch of immunity suppressed in the infant. The immune system is not brought into balance. Similar results have been found by other researchers measuring immune hormones and other factors needed to help with Th1 immune responses. Measures of airway inflammation are also increased in cesarean versus vaginally delivered children. These studies indicate that the infant immune system is imbalanced in cesarean-delivered children, and that a higher level of tissue inflammation is likely during certain environment exposures.

As expected with childhood immune disorders promoted by cesarean delivery, NCDs linked with immune problems occur more frequently in these children and adults. A study out of Denmark examined two million children born between 1977 and 2012 for possible birth-delivery-mode-associated disease. After correcting for other factors, they found that asthma, systemic connective tissue disorders, juvenile arthritis, inflammatory bowel disease, immune deficiencies, and leukemia occurred more frequently in cesarean-delivered children than in those from vaginal births. Not surprisingly, with this extra disease burden, C-section-delivered children were hospitalized more often than vaginally delivered children. The researchers suggested that a common immune mechanism likely exists in this case. Other studies have reported a variety of NCDs to be at an elevated risk with cesarean delivery. These include obesity, autism spectrum disorders and ADHD, high blood pressure, celiac disease, IgE-mediated risk of food allergies, and atopic dermatitis. It should be noted that some of these disease associations involved other factors as well. For example, with atopic dermatitis, it was the combination of cesarean delivery, antibiotic use, and certain immune gene variants that produced a significantly elevated risk of disease. Also, with celiac disease, a host immune genetic component affects who is at greatest risk from cesarean delivery. That suggests that not all babies are at exactly the same risk with cesarean deliveries for a given NCD. It may explain, in part, why different diseases show up in different cesarean-delivered children. Finally, there is elevated risk for multiple NCDs linked with C-sections. However, that does not mean that a specific child will develop an NCD if born via cesarean. It simply means that, as a group, more cesarean-delivered babies will develop significantly more chronic diseases as they age. When it is your child with one or more of these diseases, the population statistics blur.

The combined evidence of mother-child microbial transfers, infant immune maturation, and risk of later-life NCDs suggests that cesarean delivery, when elective and not medically necessary, is to be avoided.

5. Misdirected Efforts at Human Safety

Attempts to protect human health, though well-intentioned, have often gone awry. With regulatory agencies and safety testing regulations fully in place, how is it that:

1. The drug thalidomide was given to thousands of pregnant women for morning sickness and thought to be safe, only to be

withdrawn from the market later after producing massive numbers of serious birth defects?

2. Asbestos was thought to be a wonderfully safe insulation material and was installed in a majority of commercial buildings and many homes, only to be removed and remediated later as a major health hazard?

3. Safety levels for lead were set, only for it to be discovered later that apparently safe levels were reducing IQs and damaging the immune systems of exposed children?

4. Bisphenol A and phthalates were included in thousands of plastic products such as baby bottles, only to be recalled later and banned in some countries because of endocrine disruption and toxicity affecting numerous physiological systems?

5. Flame-retardant chemicals such as polybrominated diphenyl ethers (PBDEs) were included in children's pajamas and furniture as a new improved safety measure only to be recalled later because the chemicals produce multiorgan toxicity and cancer?

The problem with the chemicals and drugs being extensively introduced, then decades later withdrawn once safety levels were reevaluated, is that millions of pregnant women and children end up exposed to unsafe, NCD-promoting levels of chemicals and drugs. Carl Cranor details these and other problems in his book *Legally Poisoned*.

Why is safety testing off-kilter, leaving massive gaps in the protection of human health? For one, you need only compare the difference between the ingenuity, creativity, and investment in new drug development with the application of science to safety evaluation to see a problem. Research to discover new drugs uses state-of-the-art science, looking anywhere and everywhere for new health solutions. In contrast, regulated safety assessment of chemicals and drugs requires the lengthy building of consensus among all stakeholders, including the pharmaceutical and chemical industries. It is glacial-speed bureaucracy at its most tedious. Nothing moves fast, and by the time any consensus is actually reached, the issue under consideration may be a decade old and no longer even relevant. Not surprisingly, it is old biology all the way. There is a huge momentum toward maintaining the status quo, such that changes in required testing protocols are more the exception than the rule.

Here is a mind-numbing example related to my own area of work. Among many new drugs termed "biologics" used to treat NCDs, some are designed to correct specific imbalances in immune hormones (cytokines) connected with NCDs. The presence of the cytokine imbalance can reflect disease status, and clinicians will track cytokine levels as a way to monitor the effectiveness of treatments with some biologic drugs. So, actual human patients were already being given cytokine drugs to change cytokine levels and better manage some NCDs. However, when it came to measuring the exact same cytokines and using those measurements to determine if new drugs or chemicals might be a health concern and cause detrimental immune changes, the same cytokine measurements were deemed too new, and their relevance to immune status too uncertain, to be used to determine safety. I remember my response at the time was "You can't have it both ways." If it is good enough to dictate when you inject cytokines into people for therapy, it is good enough to tell you about the status of the immune system. But of course that requires the same level of science be applied to drug safety as is applied to drug development and therapy.

That is only one issue explaining how we could expose generations to chemicals and drugs that have the potential to cause NCDs before yanking them from the market after problems develop. The second challenge of misdirected safety testing concerns the microbiome being our filter and gatekeeper. All of our required human safety testing, to date, has been based on the model that we are only human mammals, and safety for our mammalian cells and tissues is the only concern. Under the new biology this limitation is a problem. We are only evaluating safety for a minority component of us. The reality is that human safety testing that does not take into account our microbiome does not adequately protect us. Things that we previously considered safe may not be if they are harmful for our partner microbes.

A team of researchers recently used a commonly employed laboratory mouse strain (not unlike those used in safety testing) to ask about the safety of a type of very common food additive: food emulsifiers. Emulsifiers are used to make food thick and smooth. After all, you would not want lumpy ice cream or gravy or sauce. In this case, the two food emulsifiers tested were among those most widely used in foods, polysorbate 80 and carboxymethylcellulose. Polysorbate 80 can be found in ice cream, chewing gum, gelatin, and food shortenings. But it is also in other products where the consumer might be exposed in several different ways. These include some vitamins, soap, shampoo, cosmetics, medicines, and vaccines. Products with carboxymethylcellulose include ice cream, laxatives, diet pills, textile sizing agents, detergents, and some artificial tears. Importantly, carboxymethylcellulose can also be used in dressings or drug-delivery systems following some surgeries. In other words, you are likely to encounter these two chemicals virtually every day of your life, and some doctors might even put them on or inside you.

These emulsifiers were found to alter gut microbe populations by thinning the mucus layer and increasing inflammation, eventually leading to inflammation-driven NCDs in the mice. The food and consumer product additives had been deemed safe based on old-biology-driven safety testing. But with information from this new-biology-based research, it appears that they are very likely to promote inflammation-driven NCDs. And by the way, that explosion of out-of-control inflammation and NCDs is exactly what has been happening globally for the past few decades. Smoothie anyone?

What else don't we know? Are products based on genetically modi-fied organisms (GMOs) safe? Have they been tested for detrimental ef-fects to the microbiome? If not, then it remains an open question. In fact, at least one recent journal article reported that the pesticide trade-named Roundup made by Monsanto has an inhibitory effect on several probiotic bacteria, including *Lactobacillus* species. Many chemicals and drugs that were previously deemed to be safe cause problems for our microbiome. We need to redo much safety testing with the new biology and the microbiome front and center. Safety testing needs to be relevant to the human superorganism.

6. Mammalian-Only Human Medicine

The way human medicine has been practiced is an important piece of the puzzle that explains why our microbial partners have been devas-tated and why we have, to this point, lost the battle against NCDs. Mod-ern medicine, as currently practiced, with its mammalian-only focus on the human patient, is the sixth cause of the NCD epidemic. To date, it has featured antibiotic overreach, C-sections promoted as a safe birth delivery mode, a misunderstanding of diet to benefit the whole human, and the application of misdirected or incomplete safety informatiom. The default has tended to be: See a doctor while sick, leave with an anti-biotic. Late-twentieth-century medicine used ideas about foods as a weight-loss tool without realizing that the food needed to be for the mi-crobes, too, and the microbes needed to be in place or the patient was likely to have weight issues regardless of dietary fads. Medical practice tended to steer clear of social choices like urbanization. After all, that is a personal choice, even if it is one that was practiced by the masses based on an incomplete understanding of health benefits and risks. But during the twentieth century, location-based medical advice was usu-ally limited to suggestions to leave a stressful job or run away from regional allergens. Medicine fully embraced cesarean delivery as a safe procedure for both mother and baby. It relied on human safety assessment that we now realize is both incomplete and often misdirected. It is incomplete because it focused only on the mammalian part of humans, and it is misdirected because it never measured the very things that are the most relevant indicators of the current epidemic of NCDs. Unless medical practices change to include the microbiome, progress on the other categories will not be enough.

If we hope to ever truly get humans healthy, we will need to shift our current model of medicine, fully embrace the biology of the human superorganism, and treat that patient.

10

PRECISION MEDICINE ENVISAGED

NCDs strike people of all walks of life in every country. Prominent people are not excluded, and many have self-identified or told stories of their family's challenges, often to bring attention to the need for cures. They include Bill Clinton and heart disease, Ronald Reagan and Alzheimer's disease, Sheryl Crow and breast cancer, Halle Berry and diabetes, Jenny McCarthy's and Drew Brees's sons and autism, Michael J. Fox and Parkinson's disease, Selena Gomez and lupus, Miley Cyrus and celiac disease, Phil Mickelson and psoriatic arthritis, Kim Kardashian and psoriasis, Montel Williams and multiple sclerosis, and Jillian Michaels and polycystic ovary syndrome (PCOS).

A scenario we see all too often in modern medicine starts with a group of women who learn that they are expecting babies that they have long hoped for. Through absolutely no fault of their own, the women happen to have NCDs. They may be overweight with diabetes and asthma or have high blood pressure and elevated risk of stroke, or heart disease, or arthritis, or thyroid disease, or celiac disease. These are common occurrences of NCDs. They seek out an OB-GYN to help manage the pregnancy and deliver a healthy baby. The NCD symptoms are managed during the pregnancy, but the diseases are not cured. The women carry microbiomes that are dysfunctional as they are aligned with one or more

NCDs. Additionally, their own mammalian chromosomes or genes have epigenetic marks that can promote NCDs. Some of these were established thanks to the dysfunctional microbiomes. Working with the OB-GYN, many of the women will have elective cesarean deliveries. Since it's a surgical procedure, antibiotics will be administered. This further compromises the microbiome. Even those who vaginally deliver will pass along the dysfunctional microbiomes linked with the mother's NCDs. The babies are delivered and appear to be absolutely fine. Job well done. It is a wonderful outcome of modern medicine. Or is it?

The babies from this group of mothers are launched on their life trajectory. They either missed getting seeded with Mom's gut microbes or were seeded with microbes connected to the NCDs. These babies are passed on to pediatricians. The frustrated pediatricians know all too well what is coming. They begin to diagnose NCDs in this group of children with seemingly little they can do to avoid it. Atopic dermatitis shows up in one at six months of age. By two years of age, several have food allergies to peanuts, dairy, eggs, or other foods. By four years of age, a few have asthma or have been diagnosed with autism. By six years of age, obesity is a concern for several and ADHD for others. At eight years of age, type 1 diabetes appears in some children and celiac disease in others. In their teens, respiratory allergies and depression are prevalent.

What do these children have in common as teenagers? They all have NCDs requiring prescription medications. In some cases multiple medications are needed. The diseases and medications are unlikely to go away. In many cases, quality of life is already impacted. What lies ahead? Current prognoses say heart disease, cancer, multiple sclerosis, lupus, inflammatory bowel disease, PCOS, and Alzheimer's disease among others—and, of course, more medications to go with each disease. These diseases rarely go away. Instead, we collect them. Finally, all of these children carry incomplete or dysfunctional microbiomes. This is precisely what our current best efforts have brought us simply because we have been treating only the minority mammalian patient. To get really serious about NCDs and whole human health, we need to take a different focus, one placed squarely on our microbial co-partners.

Having discussed the inability of health and prevention measures taken so far to resolve the NCD epidemic, we now must look for solutions. We must allow the new biology to flow into a reenvisioned kind of medicine.

It doesn't mean that every good thing we have been doing in medicine must be discarded. It does mean that everything—every medical procedure, every drug, every major therapy—needs to be scrutinized in light of the new things we have learned about basic human biology. We will benefit from blending the old with the new. There are many bright spots ahead on the path of whole human health, including more preventive care, personalized medicine, and broadened access to and sustainability of health services. The goal is reversing the NCD epidemic and allowing talented health care professionals to manage health instead of managing disease symptoms.

Precision medicine for the superorganism will treat you like an ecosystem. All of your body's thousands of species on the skin and in the gut, mouth, nose, airways, and reproductive tract need to be included within your health management. This is the future of medicine. It is one of the reasons a June 2015 CNBC report on medicine called the microbiome "Medicine's Next Frontier." This change is not something to fear or dread. In fact, it is very exciting and full of remarkable promise, even if some questions remain.

We already have a pretty good idea what this new, more precise medicine will look like. There are different strategies and routes and specific microbes that can be involved in manipulating the microbiome toward being healthier. This is possible no matter the individual's age and can help to prevent or treat NCDs.

Here are twenty major shifts I see ahead involving the microbiome and NCDs.

1. Microbiome modification (called rebiosis) will become standard and personalized to the individual, life stage, sex, and ancestry. For example, women contemplating pregnancy will be counseled by gynecologists on the importance of a complete, well-balanced microbiome and potentially evaluated for probiotic rebiosis as part of regular preventive medicine. It is important for their health and that of their future babies.

2. Pregnant women carrying NCDs and a dysfunctional microbiome while under the care of an OB-GYN will be offered microbiome adjustments for the benefit of both the pregnant woman and her baby. OB-GYNs have a new special mission and perhaps the best opportunity to affect the future health of two lives. A future mother with NCDs is a prime candidate for a microbiome makeover.

3. Every pregnancy will have a birth plan that includes delivering a completed baby, one fully seeded with a healthy, balanced microbiome. When delivery by cesarean is needed, techniques such as swabbing the baby with the mother's vaginal microbes at birth might be used. The precise techniques will change, but it won't be left to chance and the available hospital bugs to seed the microbiome. Feeding the whole baby, including the microbiome, will become a priority.

4. We all know of the importance of breast milk and the misuse of overpromoted formulas. But this, too, is an area in which we will see big changes. Prepping the breast milk for the baby (and even alternative probiotic-supplemented formulas as needed) will become a priority. We know that what happens during

pregnancy and birth can affect the microbes contained in human milk. As a result human milk is not all the same or equally helpful for the infant. More attention will be paid to specific factors that affect microbe diversity in human milk.

5. Antibiotics will no longer be administered without complementary probiotic therapy designed to restore and replenish the damage to the microbiome caused by the antibiotic. In fact, drugs will be prescribed based in part on microbiome status, and in some cases, probiotics and prebiotics may be given along with drug prescriptions to aid effectiveness and/or safety of the drug.

6. Prevention of disease will be given a higher priority. Pediatricians will check the baby's microbiome as a routine evaluation for risk of NCDs. If warranted, rebiosis with probiotics and diet modification to feed the microbiome will be recommended.

7. A greater focus will be placed on preventing comorbid NCDs to reduce the ongoing march toward more and more disease. As part of this, pediatricians will treat the microbiome at the same time they treat the very first signs of a childhood NCD (e.g., asthma, atopic dermatitis, type 1 diabetes, food allergies, and celiac disease).

8. Neurobehavioral conditions will be treated more often with combinations of probiotics and diet than with heavy-duty, lengthy-duration prescription medications. Probiotics are now viewed as delivery systems for neuroactive compounds. It is a new method of what is likely to be a safer, more specific, and more natural drug-delivery approach. For example, studies in mice found that treatment with the probiotic bacterium *Lactobacillus rhamnosus* GG is just as effective in controlling obsessive-compulsive disorder as is the drug Prozac. Recently,

probiotic supplementation in human petrochemical-industry workers was found to reduce their symptoms of depression and anxiety, suggesting that probiotic approaches might be able to replace other conventional pharmaceutical-based therapies.

9. Cancer treatments will include probiotics as a routine part of any radiation, immunological, or chemo therapies because microbiome status affects whether these treatments work.

10. Treatments of metabolic syndrome, obesity, and diabetes will include rebiosis as part of an integrated approach to create a permanent shift in metabolism, get rid of certain food cravings, and reduce inflammation in patients. Dieting isn't just about willpower anymore, if indeed it ever was.

11. Monitoring of systemic low-level inflammation will take on a greater role in patient screening, and the response to improper inflammation will be largely microbiome based. This is likely to be accomplished using the same kind of samples that are already routinely collected (e.g., serum, saliva, urine, and fecal).

12. Safety determinations of drugs and environmental chemicals such as pesticides, new medications, and food additives will consider safety of the human microbiome. In the absence of those data, appropriate precautions will be taken. I suspect this will happen first with pharmaceuticals and specific labeling, given the liability. There will be the need to avoid giving drugs to individuals with certain microbiomes where adverse outcomes are likely but could be prevented.

13. Microbiome evaluations will become part of the annual physical exam, and those records will be kept to track any red flags that could signal impending changes in health. You know (roughly) the history of your blood pressure and cholesterol,

but soon everyone will be able to monitor the history of their microbiome.

14. Dietary recommendations will be intrinsically linked with microbiome recommendations since the two are no longer practically separable in terms of predicting the nutrients that actually reach our internal tissues, organs, and cells. Doctors will be more proactive on using combined treatments (diet plus microbiome adjustments) than simply advising patients to eat a more heart-healthy diet.

15. Geriatric medicine will shift to emphasize probiotics and diet to aid nutrient absorption, reduce inflammation, and better support the integrity of tissues and internal organs. Loss of appetite, reduced absorption of nutrients, and diminished neurocognition are significant problems in older individuals. They become nutritionally deficient and may not easily recognize the problem. Restoring a robust microbiome can help with each of these limitations.

16. Prebiotics will be specifically tailored to feed newly installed microbes. For example, partially hydrolyzed fiber from the guar plant is a prebiotic that specifically stimulates the growth of *Bifidobacterium* bacteria and their production of butyrate. Feeding newly installed microbes will be a major new health food growth area that encompasses both nutrition and microbiology. The search for new prebiotics that are very specific in supporting only certain microbes will be a growth industry. This actually provides a new opportunity for Big Pharma, as combined drug-probiotics-prebiotics strategies will be an efficient way to correct NCDs. Note this kind of approach is already being used within the poultry industry to help chickens given probiotics to digest their food more efficiently.

17. Bacterial metabolites like short-chain fatty acids, tryptophan metabolites, indole metabolites, and sphingolipids—as well as larger, more complex molecules—will be useful new drugs in the fight against NCDs.

18. Breathalyzer tests will be used to collect information on the microbiome. This should become part of the annual checkup and eventually be used in some sick-visit evaluations as well.

19. Microbiomics will become a specialty within the medical field, with doctors training in the assessment of the microbiome's status for indicators of health and signals that NCDs could be in a patient's future. It will be interesting to see which medical schools take the lead in this effort.

20. Microbiome-based therapies are likely to be well received by health insurers. It is a comparatively low-cost treatment approach once analyses of the patient's microbiome become routine.

How do we know the idea of superorganism medicine isn't just some kind of fad? That was among the wide range of intriguing questions to greet me following my microbiome lecture before a record crowd of largely pharmaceutical company scientists at a recent drug discovery and safety meeting held in New Jersey's rich "pharmaceutical alley." Most of the largest drug companies in the world have corporate sites within a few miles of this meeting location, and several others call home to megacomplexes just a short distance across the border in Pennsylvania.

After all, I had just told a room full of the brightest pharma and biotech scientists (plus a few drug regulators, academics, and a former astronaut in attendance) that nothing would stay the same in medicine and human safety as major changes were coming. The old biology was out, and the human superorganism was here to stay. I told them that the basis for looking at drug efficacy and drug safety would change and that we would soon be working through, or at least in concert with, the microbiome. Much of safety testing had to be redone or reconsidered since the prior work had excluded consideration of the microbiome. Antibiotic administration would soon require complementary probiotic therapies or be considered a potential form of malpractice. The patient was not what they had always been told he or she was but instead was something quite different.

This was startling news for biomedical scientists, some of whom had years of personal blood, sweat, and tears invested in specific new drugs working their way toward final approval and/or the marketplace. They took the news very seriously. But before the attendees returned to work after the conference and immediately rewrote all of their job descriptions, it was reasonable to ask, is this just a fad? Is it the Hula-Hoop, pet rock, disco, twerking (yuck), karaoke (please tell me that is a fad), mobile phones—oh, wait, the last one is not really a fad either. The reason superorganism medicine is not a fad is that there is nothing temporary about our new biological understanding of you as a superorganism. The exact way we respond to it might be transitory, but not the fact that we have microbial co-partners, have had them for millennia, and know that they impact our health and well-being. We ignore them at our own peril.

Once I left the speaker's podium in New Jersey and returned to my seat, my conference neighbor leaned over and commented to me, "I think this is a paradigm shift." A half hour later I was on the phone with someone who had attended one of my lectures on the microbiome four days earlier and a continent apart (in California). He had almost exactly the same comment, "This seems like a paradigm shift."

The Able Medical Profession

If you think about it, change in medicine is not really that problematic. It is only the matter of degree and speed of change that can challenge doctors and patients alike. Doctor visits, hospital and outpatient care, prescription and over-the-counter drugs, medical tests, and medical monitoring devices do change every few years. You may have noticed this. A comparatively recent change is that everything is now electronic. Doctors are never without their tablet appendage or a nearby laptop. Many drugs that used to be prescription are now available over the counter. More high-tech instruments are common in more and more doctors' offices. Patient monitoring for blood sugar/insulin levels is very different from a decade ago. So change happens in medicine; nothing stays the same. But these examples have been a comparatively slow progression, comparatively narrow in scope, and mainly involve the doctor, monitoring, and drug-treatment end of things. Up until now the patient has stayed largely the same.

Medicine has no choice but to change in a major way because you as a superorganism are a very different patient. You are probably familiar with the preliminary data collected once you enter a doctor's office or hospital. They usually take a medical history (or ask about updates) and get blood pressure and temperature measurements from you. They may even order blood tests and review those results. But did they check the status of your microbiome or whether there had been any change in it since the last visit? Wouldn't it be useful if they had been tracking your microbiome status with each annual visit and change in health status? Wouldn't it be useful if they had monitored changes in your microbiome after you had been on drug therapy and could make adjustments?

In your last doctor's visit, did they prescribe a drug without knowing the status of your microbiome? Did they prescribe an antibiotic without a complementary plan to reinstall or repair your microbiome after it was damaged by the antibiotic? Or were you left much like a damaged coral reef?

The first step to applying superorganism medicine is an evaluation of the existing microbiome in a patient. That will be the benchmark for planning adjustments to the microbiome as well as considering therapeutic strategies (e.g., drug selection and dosing in light of the microbiome). Think of a patient's microbiome fingerprint as a new version of the standard blood pressure readings, temperature measurements, and blood chemistry profiles all rolled into one thing. Microbiome analysis will become basic stuff that will tell the doctor if you are 100 percent complete or are missing a few critical microbial species you need to be healthy.

At present, there are three main ways to do the microbiome analysis. Samples of skin scrapings, nose, cheek, and urogenital swabs as well as feces can be used to measure the microbial species and their relative abundance based on species identification. This is known as a taxonomic approach. But sometimes subsets of the same species of microbes can carry slight but meaningful differences in genes. Therefore, an evaluation of the microbial genes that you carry and their abundance can be useful. This is called a metagenomics approach. The term refers to analysis of the genomes from a community, and that is exactly what is done when a sample of your microbiome is being analyzed in this way. Finally, the end information a doctor may need concerns the chemicals that your microbes are making. It is much like your blood chemistry profiles except that this is very detailed and shows a fingerprint of the metabolites produced by your microbiome. This is called a metabolomic analysis. All three are useful.

With this as a baseline, it is now possible to tailor treatments designed to adjust the microbiome. By having individual patient information on the microbes present, the treatments follow the push to have personalized medicine using precise adjustments. From there, treatments all have one goal in common—to adjust part or most of the microbiome depending upon the circumstances. These adjustments are called rebiosis. Basically, rebiosis involves installing a good balance of microbes in the gut or in other body locations (e.g., vagina, skin, mouth, airways) and feeding the ones you want to keep around. That means feeding them what they can best use as food to survive and thrive. It is a microbiome makeover. This will not replace or eliminate the medical options that exist. But in the end, it will make those medical therapies more effective and less dangerous.

Changing the mix of microbes in a given body site can change our metabolism, physiology, and immune status and break the stranglehold of certain NCDs and/or prevent the emergence of these diseases. The challenge is to match the right microbiome alteration with the disease and patient, and that is where a broadening knowledge of specific microbiome metabolism will be useful. In Part Three of this book, I will discuss the full range of microbiome-modification strategies, but the type of applications already being used or that appear promising include new strategies for disease prevention, complementary therapies to existing treatments, stand-alone new therapies, and major microbiome reconstruction.

Antibiotic administration may kill pathogenic bacteria, but treatment can also leave the patient vulnerable both to recurrent infections and NCDs connected to a depletion of the microbiome. Many people have had the experience of getting sick a few weeks after their primary infection was eliminated by antibiotics. By administering specific probiotics, often with their preferred food (prebiotics), some antibiotic regimes can be made more effective by reducing the risk of unintended microbial damage and later health complications. This was the finding in a study done in Italy that combined antibiotic therapy to treat prostatitis with a mix of probiotic microbes (called VSL#3). With the combined treatment of antibiotics and probiotics, patients had significantly fewer complications.

Another strategy is to administer specific microbiome components to produce particular types of infections and, in the process, shift the immune response and/or better control inflammation. One of the ways this has been used is to address allergies. Research MDs have investigated gut microbe modification through the use of parasites, specifically parasitic worms called helminths. The helminths make certain chemicals that alter our capacity to tolerate external environmental factors (including worms—unpleasant as they nevertheless are). The idea has been around for a while in treating allergic diseases, but the full understanding of exactly how it works has only recently emerged. Helminth worms don't actually suppress the immune system. They simply modulate the part of our immune system that protects the helminths from aggressive immune attack. The worms are doing nothing more than protecting themselves. However, in doing so, they dampen down the exact type of responses and over-the-top inflammation that produce allergies. The treatment is not without some controversy. But it shows the power of microbiome manipulation in applying the new biology to medicine. Perhaps this knowledge will lead to a less controversial therapy.

You don't have to have antibiotic therapy to enjoy health benefits from probiotics. Probiotic supplementation can be used as a stand-alone therapeutic strategy. One of the better-studied mixes of probiotics is named VSL#3. It was recently tested in several NCD clinical trials. In a study in India of patients with liver cirrhosis, VSL#3 was administered daily for six months. At the end of six months, the probiotic group had significant reduction in both hospitalization and liver disease scores compared with controls.

A research group at the University of Kentucky conducted a meta-analysis of five separate clinical trials examining the effects of VSL#3 on ulcerative colitis. Most of the patients had the probiotic along with conventional therapy versus controls who had the conventional therapy alone with no probiotic. Remission rates for the disease in the experimental group were almost double that of the conventional therapy alone (43.8 percent versus 24.8 percent). The probiotic mix significantly improved the treatment outcome for this autoimmune disease. In other clinical trials, VSL#3 produced useful changes in studies of heart disease, nonalcoholic liver disease in children, and irritable bowel syndrome. In the last case, it was suggested that VSL#3 increased melatonin levels, and that may have contributed to the improvement in symptoms.

In some cases, even a single strain of bacteria given at the right time of development appears able to shift important maturational processes going on in the immune system. One example concerns the gut bacterium *Lactobacillus rhamnosus* GG and risk of food allergy. This bacterium appears to tip the scales in favor of a healthier immune balance and oral tolerance to cow's milk. It can prevent an overabundance of Th2-driven responses to food allergens by shifting the way the immune system interacts with foods. Additionally, this probiotic bacterium has been given in pediatric clinical trials in Australia along with oral immunotherapy for peanut allergy. The combined treatment with probiotics produced success in 82 percent of the treated children (compared with only 3.6 percent of controls). This shows the potential for treatment with even a single type of probiotic bacteria to counteract infant immune problems that promote NCDs. Maybe peanuts won't be forever the bane of parents' lives.

In another study, patients with rheumatoid arthritis were treated for eight weeks with probiotic supplements containing *Lactobacillus casei* versus controls. The probiotic-administered group not only had a significant reduction in traditional symptom scores for this disease but also had decreased levels of three pro-inflammatory immune hormones. The latter observation suggests that the bacterium was able to ramp down or resolve the inflammation that had been supporting the arthritis.

Key microbes can be used either as targets of new therapies or as biomarkers to measure the progress of existing therapies. One example of this is the previously mentioned bacterium *Akkermansia muciniphila*. It turns out that the prevalence of *Akkermansia muciniphila* bacteria in the gut is a useful indicator of dietary manipulations that were effective on overweight/obese adults. Researchers noted that if they saw no changes in *Akkermansia muciniphila*, the diets were not going to work. Of course this also suggests that simply changing the prevalence of *Akkermansia muciniphila* in the easiest possible way might be a useful weight-loss strategy.

When the microbiome requires major changes, complete reconstruction can be performed. Such was the case of Grant Fisher from Wisconsin. As an infant he was deathly ill, having developed bronchitis at ten months old. A standard course of antibiotics cleared his airways, but he soon developed disturbing GI tract symptoms. He lost weight and was not thriving due to a *Clostridium difficile* (*C. diff*) infection. The doctors gave him more antibiotics in combination but to no avail. The little boy was dying. By eighteen months, Grant was on his deathbed. Then the doctors tried a radical strategy.

Knowing that such huge, prolonged courses of antibiotics kill off friendly as well as harmful gut bacteria, they performed a fecal microbiota transplant (FMT). They took some of his mother's stool and transplanted it into his gut. It worked like a miracle.

Within twenty-four hours, Grant's symptoms had disappeared. Within a week, tests could no longer detect *C. diff* in his system. The transplant swamped out the pathogen and reestablished a healthy gut microbiome. Grant's life was saved. Agriculture has been using similar strategies for more than forty years to protect against problematic microbes.

The FMT procedure itself has gotten technologically better during its brief history. Originally, the transplant was given via a type of colonoscopy procedure. The Mayo Clinic in Arizona has used the procedure for several years for recurrent *C. diff* and reports a more than 90 percent cure rate. Recently, FMT has been successfully performed with frozen capsules taken orally. It is a less risky procedure.

Is FMT usable to treat any other diseases? The answer appears to be yes, but the exact range of its utility is still debated. Of course one of the wild cards in the procedure is the donated poop. It really needs to be from a healthy person with a well-balanced microbiome. Otherwise, you might transplant a dysfunctional microbiome and set up the recipient for other diseases (just like in the case where the mice became obese after receiving a transplant of obesity-associated gut microbes).

FMT appears to be useful for other gastrointestinal disorders. In particular it has been successful in treating ulcerative colitis (UC), one of the inflammatory bowel diseases. While not the 90 percent cure seen in *C. diff* infection, FMT has been 25 percent successful in ulcerative colitis patients. That is still remarkable progress in attacking this disease compared with the previous alternatives, which mainly employed immunosuppression. The investigators suspect the success rate can be increased once key microbes that are needed in order to reverse UC can be selected for inclusion within the donated poop.

A question remains about FMT for treatment of NCDs outside the gut. More trials are needed, but the biology suggests that it should be useful if the right donor is employed. One of the areas of investigation is with metabolic syndrome. Insulin resistance is a prelude to diabetes, and some studies have looked at FMT and its effect on insulin responsiveness. Because researchers know what some of the missing microbial signals are that prevent the control of obesity and diabetes, it may well turn out that targeted transplants or specific probiotic mixes are all that are needed. One of the initiatives under way is to standardize donor microbiomes since this is a variable that changes significantly among various clinical studies.

Finally, there is reason to believe that microbiome-based treatment of some conditions may not even require live probiotic bacteria or FMT. Several research groups have isolated gut microbial chemicals that can modulate the immune system, resolve inflammation, and change other physiological systems. These fall into several different categories, including sugars, fatty acids, and lipids. For example, Harvard microbiologist Dennis Kasper had encouraging results using polysaccharide A, a component of the bacterium *Bacteroides fragilis.* This bacterial sugar can dampen the immune system in cases of autoimmune disease and holds promise for treating diseases such as multiple sclerosis. Changing the balance of short-chain fatty acids (SCFAs), which are a fermentation product of specific gut bacteria, can have profound effects on both the immune system and the brain. A number of clinical research groups are pursuing using these chemicals in corrective therapies of NCDs with a particular focus on neurodevelopmental and neurodegenerative conditions. It has been exciting to discover that sphingolipids produced by gut bacteria can improve brain function and prevent dementia. This is potentially another whole category of microbiome-based therapeutics.

The research and clinical findings to date clearly show the value of focusing on the microbiome, in particular regarding the prevention and treatment of NCDs. If modifications to the microbiome are made, and balance and completeness is obtained, and yet the disease still remains, then the same standard drug therapies for symptom management (e.g., statins, antidepressants) with their sometimes severe side effects are still available. But to administer these drugs first without ever addressing a dysfunctional microbiome dooms the patient to a high probability of more physiological dysfunction and additional NCDs later in life. Our best health care providers will not soon forget the lesson of digoxin and food emulsifiers when it comes to the importance of the microbiome. Nor will they forget what we now know to be the largest part of our biology.

But what are nonprofessionals doing as this new paradigm emerges? What can we ourselves do with this new knowledge? Let's turn to Part Three.

PART THREE

CARING FOR YOURSELF

11

YOU, THE VOLATILE ORGANIC COMPOUND

Compared to images of attractive people on video screens, in glossy magazines, and even stories about them in good old-fashioned books, you are smelly. This is because you are substantially a volatile organic compound. Perhaps the best first step toward understanding your own microbiome and even doing something about it is to recognize this fact.

I've emphasized the importance of microbial metabolites or chemical by-products. Our microbes are busy enough making them to fulfill many a chemist's dreams. Microbial metabolites include sugars, fatty acids, and lipid compounds as well as alcohols, ketones, aldehydes, and even smelly gases like sulfide and methane. Yes, it's true. Humans make methane gas just like cows do.

The amount and variety of microbial metabolites is truly impressive. But given that microbes are a majority of our makeup, it's not surprising. Those metabolites, along with our energy sources, affect our overall chemical makeup.

Some of the chemicals we make are structurally designed to build our cells, organs, tendons, muscles, and bones. However, many are smaller molecules exuded into the gases and liquids that come from our body (such as tears and urine). Other chemicals waft off our skin into the surrounding air. These are usually found in sweat and are what necessitate deodorants for many of us. These chemicals easily enter the air because of a property called high vapor pressure and are known as volatile organic compounds, or VOCs.

You may have heard of industrial VOCs, such as formaldehyde, before. However, more VOCs are produced by plants and microbes and are largely harmless. Many, though not all, VOCs carry a scent the odor receptors in our nose and olfactory glands recognize. An example of a microbial scent I am sure you are familiar with is the mildewy, musty smell found in the bathrooms, showers, and even tents of public campgrounds. This smell is caused by the chemical tribromoanisole.

People say you are "cutting the cheese" when someone farts. However, an even closer cheesy odor, at least to really smelly ones like Limburger, is actual foot odor. One of the sources of foot odor is the bacterium *Brevibacterium linens*. It makes the VOC called S-methyl thioester. Brevibacteria produce that chemical from the breakdown of fatty acids and certain amino acids.

The VOCs, including the smelly ones, have certain functions. Some of them can aid communication among organisms, including among microbes as well as between microbes and humans. Others help to control the balance of microbes in places like the gut. However, it is doubtful they are produced solely so we can smell them. It's more likely we found it useful to be able to detect certain chemicals with our noses, either to avoid them or to gravitate toward them. It is hard to stop yourself from pausing to breathe in some more of that honeysuckle creeper as you pass by a front porch, or keep up your shopping cart pace as you trundle down the aisle with all the chocolate and confectionaries. Chocolate has about a thousand different VOCs.

Beyond promoting states of health, microbially produced chemicals also have commercial value. They are used as sources of perfumes and flavorings. They can even be engineered to make fragrances mimicking those from rare plants. Food scientists are trying to figure out which of those thousand VOCs in chocolate are the ones we really can't resist. The microbially produced VOCs are more often useful in creating perfumes, whereas the non-VOC chemicals microbes produce are frequently used to create flavors that combine both aroma and taste.

The same perfume smells differently on different people. How does that happen? Our microbes churn out a major part of what gives us our own distinctive aroma. When this combines with the scent of a perfume, the odor shifts. The difference in each individual's skin microbiome combines uniquely with each perfume, concocting a whole new personalized fragrance blend that other people can readily detect. That is why the blend of a given perfume on you is distinctive and different from the aroma of that same perfume worn by your neighbor.

Butyrate (also called butyric acid) is one of the most important microbial chemicals when it comes to signaling in the brain and the immune system. It is also a bacterial chemical you can smell. In its purest form, it smells like human vomit and is actually a major constituent of it. Once slightly modified, butyrate is used commercially to give foods a pineapple flavor and aroma. Another microbial product is propionate, and the balance of the production of butyrate versus propionate in the gut is important in certain NCDs. Under some circumstances, butyrate can be protective. For instance, the levels of butyrate-producing microbes are higher in people with healthy guts and very low in people who have IBD. Taking probiotics high in butyrate-producing bacteria helps to repair the epithelial barrier. Being able to detect levels of butyrate could be useful in balancing gut microbes and their chemical metabolites.

The ability to detect microbial metabolites as a way to evaluate your microbiome is becoming increasingly important. Being able to do so through the sense of smell has some advantages. For butyrate, humans can detect a concentration of ten parts per million. But we are rank amateurs at odor detection compared to some animals.

Furry Microbiome Detectors

Some animal behaviors seem downright rude to humans. Ever watch dogs greet one another? If they are anything like our two dogs, they smell body parts, the butt in particular. Working at Cornell University's Veterinary College, I get to observe this behavior a lot as animals come into the clinic. I used to attempt to stop my dogs from doing it whenever possible. Now I know better. If a dog is minority canine and majority microbial by cell numbers, what do you think they're picking up?

The butt and other orifices are the open portals to the microbiome. A dog will go so far as to stick its nose in another dog's poop. Why is it doing that? It is asking basic questions about the other animal like "Who are you?" and "How is your health?" And dogs aren't limited to other dogs. They can pick up this information about other animals and even humans. Their nose knows.

According to Simon Gadbois and Catherine Reeve of Dalhousie University, the social network of dogs includes pee-mails and nosebooks. Dogs' keen sense of smell makes them valuable in many ways. They can make very subtle distinctions about odors. This makes them valuable for sniffing out bombs and illegal drugs, and finding the track of lost persons, including those trapped or killed in the rubble of disaster sites.

More recently, dogs are being trained as medical service animals. A Diabetic Alert Dog can detect changes in blood chemistry signaling impending hypoglycemia in time for its owner to take action. The person is then able to get treatment and avoid a life-threatening situation. Other medical service dogs can detect colon cancer from breath or stool samples. They are picking up the specific chemical signature of the disease.

Dogs also have the capacity to make fine distinctions in cancers. They can distinguish between lung and breast cancers, detect different forms of ovarian cancers, and identify bladder cancer. The full range of talents a medical service dog could employ is not yet known. They may pick up cues beyond scent that clue them in to an owner's impending emergency.

Now dogs are being trained to detect microbes. While we can only pick up butyrate at ten parts per million, dogs detect it at thousandfold lower concentrations. If humans are scent-detecting amateurs, dogs are professional sniffers, a trait that makes them very valuable.

One of the reasons dogs and other animals are able to detect specific microbes and their by-products is that microbes consume specific nutrients and excrete specific chemicals. All bacteria, archaea, and microbial eukaryotes have their own unique profiles of excreted chemicals, some so specialized they equate to microbial fingerprints. Biologists and chemists call these fingerprints biomarkers, which are signs that a particular microbe is present at sufficient levels to be detected. And dogs are one type of animal that can be easily trained to consciously pick up these microbial chemical signatures.

One of the first uses of dogs trained in microbial scent detection was to pick up the odor of microbial growth such as that of bacteria and mold in buildings. In 2002, researchers at the National Public Health Institute of Finland demonstrated that dogs could be trained to detect bacteria, as well as strains of mold, based on scent.

A beagle from the Netherlands named Cliff knows the human microbiome so well that he can direct doctors as to when medical treatment is needed for patients. He's on staff at a hospital and even has his own uniform to wear to work. While other dogs can detect scents, Cliff's scent receptors are so keen and his training so good that he can detect changes in a single type of gut microbe, *Clostridium difficile,* a gut pathogen. Cliff can identify which patients carry *C. diff* and which don't, and he has even been able to detect impending outbreaks of *C. diff* up to three days before other instruments can detect it. This means he is able to warn hospital staff so they can take action to ward off a full-blown outbreak.

Even when humans aren't present, dogs can still pick up microbial odors. Although it has not yet been proven, researchers have suggested that tracking dogs used in searching for lost children, making mountain rescues, or finding escaped prisoners by following their scent probably use the fragrant combo of our microbes, dead skin, and oils to track people over long distances. In California water-quality projects, trained dogs have been used to determine if surface and drain water sites have been contaminated by human fecal matter as part of a prioritization program for water-quality remediation. When a water supply is suspected of being contaminated with human waste, dogs can pick up the scent of the fecal microbes faster and with greater sensitivity than can other field tests. Dogs can then easily track the contamination back to its source in order to help officials correct the problem.

In case you thought dogs were alone in their microbial scent detection, giant pouched rats have been trained to detect the bacterium that causes tuberculosis. Personally, I would prefer Cliff to screen me rather than a giant pouched rat, but to each their own.

The Scented Landscape

Virtually all parts of our microbiome play some role in our own odors and, to some extent, our individual smell and taste. Microbes in our gut, mouth, skin, and urogenital tract are major players in this. In the human superorganism, the emitted air and excretions from these sites can say a lot about the status of our microbiome as well as our potential health risks.

It could be said that Michael D. Levitt, MD, made his career in gas exploration. Levitt served at the Minneapolis Veterans Affairs Medical Center and as a professor in the department of medicine at the University of Minnesota Medical School. In 2006, the *Annals of Improbable Research* chronicled the progression of forty years of Levitt's fart-focused papers with titles such as "Studies of a Flatulent Patient" (*New England Journal of Medicine*), "Flatulence" (*Annual Review of Medicine*), and "Only the Nose Knows" (*Gastroenterology*), moving on to "Evaluation of an Extremely Flatulent Patient" (*American Journal of Gastroenterology*). With the discovery of the human superorganism, scientists and laypeople can appreciate his observations of the gut microbiome.

Production of gas itself is not necessarily a bad thing. In fact, some gas after eating some types of food (e.g., fiber-rich food) is perfectly expected and a sign that particular microbes are metabolizing those foods. Among these microbes are the ancient archaea that evolved over a billion years. They are separate from the branch that mammals like us grew from, but they are also separate from bacteria. In many ways they are a bit of a hybrid, looking and acting like bacteria but possessing a lot of cell machinery that is more like what we have. Archaea produce gas in marshlands, they do this in cows, and they do it in our own guts.

While many gases produced are odorless, the end production of sulfur is definitely what gives gas its repulsive odor. But sulfur-containing compounds such as the sulforaphane found in broccoli, Brussels sprouts, kale, mustard greens, and cabbage also have reported anticancer qualities. The same chemical has been studied for its apparent benefits to patients with autism spectrum disorder.

The skin is a remarkable and expansive site for our microbial co-partners. Research on the skin microbiome has been slower to develop compared with that on the gut microbiome. However, recent findings indicate that skin microbiome manipulations are also going to offer a wealth of possibilities as we move toward superorganism medicine. Our skin covers us from the top of our head to the bottom of our feet and everything in between. It has many different local habitats for microbes, ranging from moist tropical rain forests to desert oil fields. Each different area has its own mix of microbes that are attuned to their preferred food sources and produce metabolites that support and modify our own body regions. You would not want the particular mix of microbes living between your toes or in your underarms to show up on your face. They wouldn't like the outcome, and neither would you.

Human skin is a major source of body odor. That is one reason it is given so much attention in terms of personal hygiene, and a lot of money is spent on myriad personal care products that enable people to become masters of their own body odor. But you might want to think about other ways to control body odor, ways that create better outcomes for your microbiome. We may not realize it, but human sweat in its pure form is completely odorless. In fact, that was established back in the 1950s. If our sweat has an aroma, it is from the chemicals made by our skin microbiome. As I will discuss, each person's aroma is a type of carrying card. We move around surrounded by what Jack Gilbert of Argonne National Laboratory has referred to as your personal microbial cloud. I like to think of it as the cloud of dirt that surrounds Pigpen from the comic strip *Peanuts* wherever he goes. We carry our own personal microbial cloud with us to work, on trips, and as we interact with one another. Other humans and animals notice our aromas, and mosquitoes zero in on them. Often mosquitoes are just a nuisance until you run into ones transmitting disease.

At summer gatherings of family or friends, some people are constantly bothered by mosquitoes, while others seem to have little trouble. Some people just seem to be mosquito magnets. These people can attract the attention of virtually all the feeding mosquitoes in the area. Foods people eat can be one factor in attracting or repelling microbes. For example, mosquitoes react to organosulfur chemicals in garlic as if they were vampires. They hate it. So garlic breath repels more than just other people.

But a major reason mosquitoes either stalk or run away from certain people is their skin microbiome. While mosquitoes can use many clues, the biggest factor in their selection of a person for feeding are the VOCs from the skin microbiome. In a study of forty-eight individuals with different skin microbiomes, researchers led by a group from Wageningen

University in the Netherlands analyzed which features of the skin micro-biome made humans most attractive to mosquitoes, and which features allowed us to hide in plain sight. They found that having a more diverse skin microbiome tended to protect you from the mosquitoes. Also, which specific bacteria were present in high numbers made a difference. Individuals with a high abundance of *Staphylococcus* bacteria were very attractive to mosquitoes, while people with higher levels of two other types of bacteria, *Pseudomonas* and *Variovorax,* were unattractive. It all comes down to the chemicals our co-partner microbes produce. Certainly, these findings may give new meaning to what constitutes truly natural mosquito repellants.

Recently, a large collaborative research group spanning the United States and Europe provided body-region-specific details of the skin microbiome. They created maps similar to those produced to depict the vegetation of the US. These maps have topography and were in effect three-dimensional. They highlighted differences between the skin microbiota in certain regions of the body (e.g., groin, tops and bottoms of feet, armpits, scalp, neck, and face) as well as differences between men and women. They also illustrated how use of personal care products can affect our skin microbiome. In addition to maps of microbial species inhabiting different areas of skin across the body, the researchers developed chemical maps of microbial metabolites showing what areas of the skin are rich in specific chemicals made by our co-partners. Many of these chemicals contribute to our body odor as they mix with our own proteins and oils on the skin. The take-home message is that each specific region of our skin has its own unique mix of microbes because the local "environment" is different. Also, each skin region is its own little perfume factory through the combination of odors emanating from hundreds of specific microbes in that skin region mixing along with our own various oils and gland secretions. The House of Chanel has nothing on us in terms of scent diversity and complexity.

To get deeper into the skin microbiome, it is useful to understand the nature of our secreting glands. There are basically two general types: sweat glands and sebaceous glands (concentrated in the scalp). Sweat glands are further subdivided in two types: eccrine (water, protein components, salt, urea, lactic acid) and apocrine (pheromones, proteins, and, importantly, more fat-loving chemicals).

The foot also has its own microbiome that differs by region. Analysis of foot odor has provided several interesting findings. It turns out that smelly feet are mainly the result of one type of bacteria, *Salmonella*, although some other less prominent bacteria can also produce odiferous metabolites. *Salmonella* inhabits the bottom side of our feet, accounting for more than 90 percent of all the bacteria present in that location. In contrast, the upper side of the foot is much more diverse and effectively "less smelly." Some combination of features of the bottoms of our feet is likely to give the *Salmonella* its preferred home. *Salmonella* metabolizes peptides into a compound called isovaleric acid. It is abundant on the bottoms of our feet and largely absent from the tops of feet. This chemical is probably the most important in giving our well-exercised feet their distinctive acidic, cheese-like odor.

Research into the microbial origins of body odors has led to comparisons for sex, age, condition, and body location. Employing a strategy called competitive exclusion, where probiotic bacteria are used to swamp out other less desirable bacteria, proposals have been made to use replacement bacteria in specific body locations (e.g., underarms) as natural deodorants. We can make artificial sweat.

The mouth is a very complex microbial world of its own. Hundreds of species of bacteria, fungi, archaea, viruses, and protozoa interact not only with you but also with one another. In the mouth, they metabolize food and anything else that goes into your mouth, signal one another and your immune system, and organize themselves as a form of complex multicultural communities. They both respond to and help to create the local environment of your mouth. This is important not just for dental health but also for two of your senses: smell and taste. The microbiome of the mouth can influence both your own sense of smell and taste and the odors in your breath. In fact, there is a chemical signature for halitosis, and some companies produce instruments for quantifying halitosis (caused by your oral microbes). It is one way to detect when our microbes get out of balance, as happens with NCDs.

You have probably heard of the aroma or bouquet of various wines. But what is interesting is that most grape chemicals are odorless. It is only after they have been acted on by bacteria that the odor-producing chemicals are created. A recent study showed that the bacteria responsible for the production of aroma from white wine grapes are located in the human mouth. The grapes have a precursor chemical that is odorless but is turned into odor-producing chemicals by mouth bacteria. It does make one wonder whether the special "palate" of wine connoisseurs is more about their mouth bacteria than about their mammalian-based talents.

Obese individuals have apparent differences in mouth and saliva microbes that affect their interactions with foods, much as the aroma of white wine experienced by the taster is affected by the mouth microbiome. Significant differences have been reported between obese and non-obese individuals in both the composition of the mouth microbiome and in local metabolism. When it comes to white wine, the obesity-associated microbiome appears to have an impaired release of aromatic compounds.

In an interesting flip side to the story of dog detection of human microbes, humans have been used as trained judges to detect malodors associated with microbially related periodontal problems in both humans and dogs. In a London dental school study, odor detection was associated with bacteria that produced high levels of volatile sulfur and were also present at higher levels coinciding with periodontal disease.

The urogenital microbiome does not escape a discussion of scents and health. I previously brought up the vaginal microbiome relative to birth and microbiome seeding of the newborn baby, but there is so much more about the microbes in the vagina as well as those in the penis that have to do with odors, life cycle, sex, and health.

The vaginal microbiome in a healthy woman is dominated by *Lactobacillus* bacteria. Within the vagina there are local regional differences in microbial makeup and distribution. Changes can happen in response to age, menstrual cycle, and sexual activity. However, estrogen level appears to be a major player in many of these changes. Women with bacterial vaginosis have an altered vaginal microbiome featuring reduced amounts of *Lactobacillus* species and an increased presence of *Gardnerella* and *Prevotella* bacteria. Treatment for bacterial vaginosis leads to an increase of *Lactobacillus* species. But after several weeks there is a risk the women will revert to the disease-promoting microbe profile and possible symptom reoccurrence. We can look forward to getting to the root causes for the dysfunctional vaginal microbes, but for now we know scents from the vagina change based on (1) secretions from exocrine glands (which secrete through a duct or opening) that can be metabolized by vaginal bacteria and (2) the mix of microbes within the vagina. Not unlike with skin odors, these scents are a combination of exocrine gland secretions, sloughed-off epithelial cells, mucus, and resident microbial metabolites, which can change depending upon a variety of circumstances, including phase of the menstrual cycle. The overall changes in vaginal scents during the cycle are probably driven more by changes in the exocrine secretions and spectrum of bacterial metabolites than by changes in the bacteria themselves, since the vaginal microbiome of women is thought to remain relatively constant across the menstrual period. The exception to that is during bacterial vaginosis, when changes in the vaginal microbiome by itself seem to drive changes in scent.

Men also respond to these vaginal scent changes. An international team of researchers showed that both salivary testosterone levels and cortisol levels in men changed based on the female axilla- and vulva-derived scents linked with the phase of the menstrual cycle.

If the vagina has a microbiome, then so does the penis, albeit with less microbial complexity and in smaller numbers. While the penis is under-studied compared to the vagina, certain points have emerged. Men differ in the penis microbiome based on whether they were circumcised, their age, and their sexual activity. Their penis microbiome appears to be affected by their sexual partner's microbiome as well. Regarding research on scents and penises, there is only a little to report about odor changes. The really obvious malodorous smells (e.g., fishiness) are usually connected to an accumulation of dead epidermal cells and microbiota changes such as the accumulation of bacteria that cause bacterial vaginosis in women. For more information on more subtle, less aquatic smells from the penis, we await additional studies.

Electronic Noses

Cliff the beagle has some new competition. There has been a technology-based effort to create an odor-detection system as powerful as Cliff's nose called the electronic nose, or E-nose. It has the capability of analyzing the VOCs in human armpit odor. It turns out that your armpit microbes live on the gland secretions from your skin in that area, and the microbes' odor-producing metabolites are sufficiently unique that they can be used to identify you. A dog could do it, but so can the E-nose. Imagine a time when you are at customs and immigration at an airport and they ask you to raise your arm so they can sample your armpit for identification purposes. Relax, enjoy your stay!

The E-nose has other applications, including a capacity to distinguish among Alzheimer's disease patients, Parkinson's disease patients, and healthy humans based on chemical analysis of their breath (exhaled air). It also can be used to detect cancer. Imagine being able to monitor NCD status as well as microbiome status all with the same piece of equipment. This capability points toward the future and what is likely to become a routine screening (as noted in the previous chapter). When it comes to physicians monitoring microbiome status and the success of probiotic administration, an E-nose seems likely to end up sitting alongside ultrasound machines in doctors' offices sooner rather than later.

The Power of Scent Receptors

Scent detection between our scent receptors and microbial metabolites appears to go way beyond just odor detection by our brain. In fact, there is increasing recognition that it is not just about odor but rather our whole physiology. Jennifer Pluznick, now of the Johns Hopkins School of Medicine, has been studying scent receptors that show up in unusual locations of the body and seem to have more functions than just odor detection. What is particularly interesting is that some of the scent receptors interact with and are triggered by chemicals made by our microbes. In one case, at least one scent receptor seems to have major effects on blood pressure and risk of hypertension. Short-chain fatty acids made by our gut bacteria appear to use at least one of these scent receptors to help regulate blood pressure. Pluznick and her colleagues found that when mice were treated with a combination of antibiotics, their gut microbiome was destroyed, bacterial production of short-chain fatty acids that bind scent receptors was severely reduced, and blood pressure skyrocketed. Supplying the mice with just the bacterial short-chain fatty acids significantly reduced their blood pressure. That was the key to the blood pressure problem. Our microbiome appears to be able to control blood pressure using a rather unexpected strategy, production of VOCs that interact with scent receptors located in tissues not even remotely involved with smell.

Back in 1979 at Cornell, I remember teaching students about then-new research studies on how mice select their preferred mates based on urine odor. It was at least the more lighthearted part of my immunogenetics course and probably a welcome relief for students from some other topics of the day. Now it turns out it is not just mice that use those types of cues.

Humans smell one another, and that has a lot to do with attraction (even if we are not consciously aware of it). In her 2008 *Psychology Today* article titled "*Scents and Sensibility*," Elizabeth Svoboda explains why the chemistry, spark, or electricity we feel with our future life partner has as much to do with scents as anything else. As it does with mice, some of this seems to do with our mammalian immune self-identity genes that control things like organ transplant acceptance or rejection. These genes sit in a complex known generically as the major histocompatibility complex, or MHC. The human version of the MHC is called the human leukocyte antigen complex (HLA complex, for short). It turns out that women are drawn more to scents derived from men carrying different immune-response genes than their own. (Anyone still doubt that the immune system does more than just fight infections?) They unknowingly prefer to produce offspring with men who can help to broaden their child's immune response capabilities. That makes perfect sense and is desirable. But how does the human HLA complex or the mouse MHC equivalent affect smell-based social interactions and mate selection? Another way to ask the question: How do human immune response genes translate into odors? There are several different theories about how the HLA complex affects human scent as described in a book titled *Human Scent Evidence*. But this is where it gets really exciting and where the microbiome comes in.

It is known that at least some of our mammalian genes do affect the microbial partners we choose (or tolerate). Among the genes reported to affect our microbiome's composition and diversity are the HLA genes. We don't have to take kindly to just any old bacteria in or on our body. We have some say in choosing our microbial partners just like we do in choosing our human mates. The immune-response genes and their proteins are not known to be smelly. But they influence which smelly microbes we have in the gut and other locations. Our co-partnership with microbes is truly that. We accept them in general; then they proceed to do home renovation within our body. In effect, the smelly microbes we have are a type of surrogate for our HLA type. So when it comes to mating, loving those pungent microbes in a potential partner can probably help you conceive a more immunologically resilient child.

Armed with an increased respect for our dogs, we can anticipate encountering the electronic nose in a future doctor's office, accompanied by our personal microbial cloud wherever we go. And with new hope for relationships that pass the smell test, we are ready to probe more deeply into self-care in the age of superorganism medicine.

12

SUPERORGANISM MAKEOVER

A focus on self-care is nothing new. Obviously, people have been doing it for centuries. Our ancestors did it, and we already do it each and every day. We practice self-care when we pick food for our meals, when we choose among lifestyle activities, when we select those we spend time with, and when we find the things that bring joy to our lives and pursue them on a regular basis. There is plenty of professional help when it comes to our health and well-being, but caring for oneself is ultimately a personal matter. We automatically become both the authority and the most relevant expert as overseer of our own bodies. Sometimes we may care for ourselves with lots of information available to support our decisions and choices of activities; sometimes we operate in a vacuum of information but have to make decisions anyway. That is the beauty of self-choice and self-care. When it comes to the outcomes of caring for ourselves, we are the ones who live with those outcomes.

You don't have to change all the things you do or love, or change everything you eat or who you spend time with. These are simply new ideas and information about the array of options you have before you.

The word "makeover" itself is a fashionable new coinage, but things like having a spa day or a massage are really not that different from a visit to therapeutic hot springs of decades to centuries ago. They are all about figuring out what can refresh and renew our bodies and bring them into noticeably better balance. They are a reset button.

People have different preferred reset buttons. Maybe yours is a day at the gym, a swim in a pool or ocean, a walk, a hike, a jog, or a bike ride in nature, pursuing your favorite hobby, or maybe it is just being with friends or family or others with common interests. If you know what always allows your body to be relaxed, feel whole, and fully work for you, that is golden. It is also a frame of reference I would ask you to hold as you consider options for a superorganism makeover. Things you can do involving your microbial co-partners need to feel helpful and be useful for you—not something that is only theoretically helpful but, instead, something that you notice in your own body as useful.

I have my own frame of reference for this as well. For me it is swing dancing, a hobby I took up in middle age. After a good dance workout that is both physical and social since it involves dancing with other people, my entire body relaxes and I have a fully restored feeling. I sleep better, and that is a great personal benchmark for me. Consider what benchmarks you might use that will allow you to compare the helpfulness of specific superorganism makeovers. Which ones work for you, and which ones don't? You know your own body. Let it be your guide.

Let's approach this question of self-care very gradually. Any reasonable person would have questions like these.

1. Do We Know Enough to Even Attempt Something like This?

Yes. Just pick up any grocery store checkout magazine and look for articles. They tout things like ... "Nutritional Breakthrough, Melt Fat with

Probiotics." I have one such article sitting in front of me as I write this. Of course, that is not necessarily the best way to gather health-related information.

Integrated medical therapies have been expanding on the health landscape, and certain parts of that are taking a systems-biology approach to health. That means not just focusing on a single specialty of the physiological system but looking across them for more holistic solutions. Yet even these integrated approaches still seem to be missing something. In a sense, it is only one small additional step to include the thousands of microbial species that are part of our makeup into an integrated approach in medicine and self-care. Beyond that, we do know what disease-connected microbiomes look like, and we do have tools for changing our microbiome.

Be that as it may, one of the things to keep in mind about the microbiome is that we are near the beginning of the learning curve. But that does not mean that we do nothing until we know all. If there is much we do not know about aspects of the microbiome, we do know what happens by doing nothing. That simply allows the NCD state to continue and more and more diseases to appear as we age. Most people would agree that is simply not good enough. They would prefer other options for improved well-being in their lives.

2. Is There a Single Best Superorganism Makeover Process?

No. There are several very useful recipes for a better microbiome balance in any superorganism, including you. Many of these recipes are likely to be of some use for you. The very best for you will likely depend upon your ancestry/ethnicity, sex, age, previous and current diet, early-life experiences, food allergies and intolerances, health history, and current disease burdens. This is a little like planning the ideal vacation. One person's perfect vacation would be another person's worst nightmare, and ideal vacations involve many different personal factors. It is the same with ideal microbiomes. At present, there are thought to be three different fundamental types of healthy gut microbiomes. These are called the three enterotypes, meaning they are each a prototype of a useful mix of microbes dominated by one bacterial genus: *Bacteroides, Prevotella,* or *Ruminococcus.*

The enterotypes differ based on people's ancestral origins and ancient lifestyles, including diet. For example, *Bacteroides* prefers to metabolize protein and animal fat, *Prevotella* is linked to mucin proteins and simple carbohydrates, and *Ruminococcus* prefers mucins and sugars. Mucin is a type of protein with lots of sugars (carbohydrates) attached to it. Because it carries lots of sugars, it is referred to as a heavily glycosylated protein. Mucins form gels that coat epithelial linings in our body, acting both as a barrier and as a lubricant. Some bacteria like to digest all or part of mucin; some don't.

In Korea two different gut enterotypes have been identified. One is dominated by *Bacteroides* and the other by *Prevotella.* It is as if these ancestors and their co-partner microbes became a well-tuned completed self a long time ago using the diets that were locally available. Scientific debate abounds as to whether there are more than three enterotypes and whether they exist on a continuum instead of as discrete clusters of gut microbes. There is much more to say on this topic, but for now, suffice it to say that you should aim for the healthiest version of the nearest one of these enterotypes to you. People were healthy with them in the past, and we can be yet again.

3. Can Your Microbiome Be Analyzed?

Yes. Complete analysis of every possible body site is yet to come, but several companies, some university-connected, offer fecal, skin, and other site analyses of your microbial mix. That can give you a starting point for your personal makeover. However, most of the services offered, to date, are based on the types of microbes, mainly bacteria, in your samples and not on their gene composition or metabolic capabilities. Personal microbiome analyses are likely to become more widely available as they move quickly from the research lab to service companies.

Keep in mind that if you have a baseline analysis of your microbiome, that can be used as a starting point to monitor changes in the microbial mix as you pursue microbiome modifications. Obviously, how your own body feels and performs is the ultimate test of whether a specific rebiosis approach was useful for you. But the microbial analysis can tell you what changed along with your improving health.

Fecal microbiome analysis is the primary way that disease associations were established involving the microbiome, the immune system, the brain, and regulation of inflammation. We can see when there are major problems. We also can see when modifications have been made and evaluate those for their usefulness. But this kind of analysis doesn't come close to telling us all we want to know. Besides the fact that it tells us little to nothing about the microbes in other body sites, such as the skin, reproductive tract, and lungs, it doesn't even tell us about all of the microbes within the entire gastrointestinal tract.

4. Do Probiotics Work?

In general, yes. To begin with, probiotics are microbes—sometimes individual bacteria, but often mixtures—that when administered correctly can improve health. There have been numerous studies with probiotics, but one of the most powerful in terms of numbers of humans evaluated was a collective analysis of all the published individual human trials that have been performed using probiotics examining gastrointestinal disease. Marina Ritchie and Tamara Romanuk of Dalhousie University in Halifax, Nova Scotia, Canada, conducted a meta-analysis using eighty-four trials involving some 10,351 patients with a focus on oral administration of probiotics. The findings were that probiotics are generally useful for both treating and preventing gastrointestinal diseases.

5. How Do Probiotics Work?

One of the best descriptions of the way this happens can be found in a consensus review article by a panel from the International Scientific Association of Probiotics and Prebiotics published in the journal *Nature Reviews Gastroenterology & Hepatology*. Colin Hill of the Alimentary Pharmabiotic Centre, University College Cork, Ireland, was the lead author on a collaborative paper with eleven researchers and clinicians located in Canada, England, Finland, France, Italy, Scotland, Spain, and the United States. The collaborators identified three types of situations in which probiotics can improve health:

1. Cases where comparatively rare microbial species perform some crucial activity that benefits the immune, neurological, and/or endocrine systems. This often occurs because the microbes make some key chemicals that affect our biology.

2. Many different species appear to produce vitamins for us, help maintain our gut barrier, neutralize toxic chemicals such as carcinogens, control bile salt metabolism, and boost certain enzyme activities.

3. Finally, many important health-promoting functions are performed by what seem to be the most prevalent microbial probiotic species in a healthy microbiome. These include competitive exclusion of pathogens (meaning that the good guys outnumber the bad guys), control of the acidity of our

body's environment, production of short-chain fatty acids that affect our cells and tissues, increase in the turnover of our enterocytes, regulation of the transit of food through our intestines, and help in bringing our gut microbiome back into balance.

6. Are All Probiotic Sources of the Same Bacterial Species Equal?

No. There are at least three ways they can differ in outcomes. First, of course, the probiotic company's production operation, especially whether it includes prebiotics, can affect what happens when you take the probiotics. Second, different source cultures of the same bacterial species (e.g., *Lactobacillus* species) might carry differences in some genes that affect the microbe's metabolism and functions. Finally, there is some evidence that the probiotic bacteria's host species of origin (say, human versus cow) might have different capabilities for signaling developing physiological systems such as the immune system. It is possible that for some types of help, human-sourced microbes are more effective in us than are microbes derived from other animals.

7. If I Am More Than Sixty Years Old, Is It Too Late for Rebiosis?

No. It is never too late to experience benefits. In fact, one of the exciting aspects of microbiome research is the potential for dramatic benefits on both ends of the life spectrum. There are many health issues of older age that are related to nutrient absorption and use as well as deterioration of multiple physiological systems, including the immune system and brain. Having a robust microbiome using more energy sources and producing metabolites that aid the immune system and brain is healthy.

8. Does It Mean I Did Something Wrong If I Need Rebiosis?

No. Rebiosis reestablishes a healthy microbiome by seeding and nourishing useful microbiota after dysbiosis or a microbial imbalance has occurred. In theory, rebiosis would be unnecessary if you were seeded at birth with a healthy microbiome and had it fed well throughout infancy, childhood, and later in life. However, even under the best of circumstances, things can happen to anyone that disrupt the balance of one's microbiota in the skin, airways, urogenital tract, or gut. Perhaps you had an incidence of food poisoning, a major stressor such as loss of a friend or family member, or an infection in one part of your body that took you out of your normal routine and forced you to change your diet. Travel on airplanes and visits to hotel rooms can bring you into contact with new microbes that can alter your personal balance. Exposure to chemicals and drugs can readily upset your microbiota. Additionally, a new NCD diagnosis—or even a flare-up of certain NCD symptoms (allergic rhinitis, asthma, Crohn's, celiac, psoriasis) and the tissue inflammation associated with it—usually results in shifts in your microbes. Rebiosis is a sensible response.

Again, this may be the most important garden you ever tend because it constitutes 90 percent of you. You will harvest what you have seeded and nourished. If you do nothing, particularly early in the growing season, weeds will grow and thrive. Harvesting almost anything from a full bed of weeds might require starting over. But with preparation, careful seeding, and early attention to the seedlings, management of the garden becomes less onerous. Minor adjustments are in order to keep the balance among plants, and weed removal is needed to produce a bountiful harvest. There is also a synergy among the plants. A row of marigolds may not produce a food crop, but their presence helps to keep down pests. Understanding the positioning and interrelationships among your garden's plants is useful. The same relationships exist among your microbiota.

9. Would Probiotics Be Helpful After Taking Medications Other Than Antibiotics?

Yes. It turns out that antibiotics are not the only medications that can alter your microbiome. Medications have not been routinely screened for safety of the microbiome, so for many, microbiome safety is likely an issue. One clue that might help sort out which drugs are most likely to cause microbiome problems would be if the medication has side effects that include GI tract symptoms. We now know that at least several of those affect the gut microbiome and can contribute to gut mucosal injury. The known list of antimicrobiome medications includes nonsteroidal anti-inflammatory drugs (NSAIDs) and proton pump inhibitors. Additionally, some drugs such as statins only work in people who carry certain gut bacteria. Probiotic and prebiotic mixes (called synbiotics) can be helpful in protecting against these adverse drug effects.

10. For a Probiotic to Work, Do I Have to See Changes in My Fecal Microbe Profile?

Not necessarily. Changing the fecal microbial profile is only one way that probiotics can affect you. They can change metabolism while they are passing through your system even if they do not take up a long-term residency in your gut. But recently Rob Knight and colleagues at the University of Colorado Boulder discussed a different type of probiotic effect. Even when commercial probiotics were consumed by humans and their communities of gut microbes did not appear to change the mix of bacteria, there were still significant changes in gut microbe gene expression that affected food preferences.

11. Are the Probiotics in Standard Yogurt Good for Your Health?

Well, yes, but maybe not in the ways you might think. Many yogurts have a limited array of probiotic bacteria, and they are not necessarily at high doses. Live-culture yogurts often contain a limited array of *Lactobacillus* bacterial species. However, there are efforts to boost the number of species represented in some yogurts, and this may help their effectiveness considerably. Where are these bacteria most crucial? The vagina is where they are normally the predominant type of bacteria. There, they make lactic acid, which acidifies the vaginal environment and prevents the growth of microbes that cause bacterial vaginosis. So ironically, regular yogurt might help your health the very best when directly applied to the reproductive tract.

12. Is the Gut the Only Target for Rebiosis?

No. Certainly more has been done with the gut in terms of research and rebiosis strategies than for other body sites. However, they are all eligible for rebiosis when microbial imbalances arise. The vagina was just mentioned relative to *Lactobacillus* rebiosis and restoration of the acidic environment. There is evidence the skin microbiome can be altered through dietary ingestion of probiotics as well as topical application. In a recent screening of 896 oral bacterial isolates, the following bacteria— *Lactobacillus crispatus* YIT 12319, *Lactobacillus fermentum* YIT 12320,

Lactobacillus gasseri YIT 12321, and *Streptococcus mitis* YIT 12322— emerged as good candidates for oral probiotics capable of inhibiting periodontal disease. For the airways and lungs, direct intervention with probiotics to modulate the lung microbiome has lagged behind efforts with other body sites. However, there is considerable interest in this area of prevention and therapeutics.

13. For Better Health, Is It Enough to Just Change My Diet or Use Probiotics Alone?

Probably not. Each individual is different, as are the circumstances. However, there is evidence suggesting that the most effective strategy is to adjust both your diet and your microbiome in a coordinated strategy. In some individuals, blunt-force changes in either diet or the microbiome have worked. But it is not easy nor is it universally effective. If it were, changing your diet would always lead to permanent weight loss and the correction of NCDs. It does not. It is very difficult for many people to lose weight and keep it off with only a change in diet. Neither does taking probiotics work well if your diet is working against supporting the microbes you are attempting to install. You need to consume a diet that allows the microbes you are installing in your gut to thrive, to have an ecological advantage in you, and to function fully. On the flip side you need microbes in place that allow you to benefit by consuming what is deemed to be a healthy diet. If you want to consume a Mediterranean-type diet but your microbes are calling for pizza and milk chocolate because they are incompatible with the healthier diet, you are in for a real internal struggle.

The use of dietary changes and probiotics went hand in hand in my own personal experience with correcting microbial dysbiosis.

———

More than a year and a half ago, I pursued my own rebiosis path with very positive health outcomes that were verified by both my family and specialist physicians. The personal encounter with microbial dysbiosis began more than thirty years ago when, around age thirty and in the final stretches of my pursuit of academic tenure and promotion at Cornell, I developed what was diagnosed as irritable bowel syndrome (IBS). The prolonged stress of pushing hard to prepare for a competitive tenure package combined with a poor diet resulted in chronic gut issues that required antacid prescriptions (now available over the counter). It took close to two years of medical intervention and the granting of my promotion with tenure to resolve the issue.

But what I did not know then was that the GI issue, and the likely gut microbial dysbiosis that accompanies it, was not finished with my system. I soon developed a pattern of recurrent sinus infections that seemed to be promoted to some extent by respiratory allergies. The infections would not resolve and eventually required antibiotics. Soon I needed three to four rounds of antibiotics every year to function, and this pattern lasted for thirty years. Resistance would develop to the routine antibiotics, and I was requiring more specialized new-generation drugs, often with a longer and more serious list of adverse side effects. The cycle continued and continued, and I was losing hope as even ear, nose, and throat (ENT) specialists seemed at a loss for a longer-term solution. In reality, there were very few days during those thirty years when I felt really healthy and full of energy. Even when I was not on antibiotics, I was often developing infections that sapped my energy and impaired both my breathing and my sleep. One can only imagine what approximately one hundred rounds of antibiotics over this interval of my adult life did in terms of the inadvertent and almost constant attack on my microbiome. Of course prior to the 2000s we really did not know much about gut microbe dysbiosis and the connection to chronic diseases and conditions.

Things got bad enough with my health that my family doctor was telling me I needed to lose weight, and my ENT specialist was arming me with more allergy-related meds, trying to stave off any potential precipitating factors for the sinusitis. The immediate response was no weight loss and no significantly improved control of the recurrent sinusitis. Then a glimmer of useful information emerged. I was able to tell the ENT specialists that around the time each sinus issue began again, I had gastrointestinal issues, and this seemed to be consistent as a connected pattern. Their response was to quiz me about gastric reflux as a precipitating factor. But I had not noticed that. It was getting bad enough that even my esophagus was irritated along with my sinuses. The ENT clinicians again focused on the gastric reflux possibility, and finally, I was awakened one night by an intense episode of reflux confirming their suspicions. What we hadn't known was that it had been happening at night during my sleep, and I had not been previously aware of it. It was what is called GERD, gastroesophageal reflux disease. That is one of those chronic NCDs.

Still, knowing that gastric reflux might be setting me up for the sinus infections was not enough to break the cycle. The gastric reflux was becoming more frequent and intense, and the soup of antibiotics continued to flow in the same pattern. In retrospect, it is no wonder that there was no significant improvement in the gastric-reflux-promoted respiratory issues. I had done nothing to correct my problematic microbiome. There was no seed-and-feed plan in place to help my much beleaguered gut microbes.

Then it happened, a totally offhand and unexpected observation showed me a way forward to address three decades of poor personal health. I traveled to Germany for a six-day, residential, scientific conference on the topic of metals and infectious diseases that I had helped to organize. At the end of the conference, we were supposed to have prepared most of a definitive monograph on the subject. During this time, we were sequestered together, working long hours, eating locally sourced, chef-prepared foods, and getting what was probably less than a normal amount of exercise. The food included local cuisine (e.g., German sausages and sauerkraut), but it also included more exotic Greek, Indian, and Chinese dishes. I ate some grains but in different forms than those I got in the United States. As usual on this type of trip, I was dreading the emergence of a sinus infection, which often cropped up connected with travel, time changes, and disrupted sleep. Just in case, I had an antibiotic refill with me from the ENT if needed.

To my astonishment, there was no sinus infection. But beyond that a remarkable thing happened. After four days of the six-day conference, my pants no longer fit me. Inexplicably, I had lost a pant size around the waist. How could this happen? I had been eating more than my fair share of what was exceptionally delicious food, three full meals a day, and getting little to no exercise. Yet my waist had shrunk! Plus, there was no gastric reflux, and I had more energy despite being jet-lagged. My loss of waist was all due to the disappearance of inflammation. Apparently, I had been experiencing constant inflammation in my body, and the change in environment in Germany had allowed that to lessen. Whatever helped to lock the inflammation in place was now absent or at least reduced.

I was convinced that something about the change in food was involved as I seriously doubted that sitting around all day and part of the night had helped my waist to shrink. I was not drinking my previously regular intake of diet soda, and that was a potential factor. Surprisingly, I ate a broader diversity of foods than normal in Germany, including some grains and some hard rolls. I immediately e-mailed my wife and let her know I had my own dietary experiment to run once I got home. I was determined to identify the factors in my US diet that kept me in a constant pro-inflammatory state. The German food had apparently aided my health, and something about my US diet had to be locking me into constant inflammation.

Of course, the conference itself provided some additional clues to my better health. It seemed that every discussion we had kept going back to the microbiome, even though that was not a specific agenda item or the topic of a background paper. We were beginning to realize that there was little point worrying about what the metal would do to the human immune system and resistance to disease if we first lacked information on what our microbes at the portals of entry (e.g., skin, gut, airways, and mouth) were doing with the metals. Every discussion session ended up being unintentionally rerouted toward the importance of learning, first and foremost, what our microbiome did with metals. For me, it seemed like the universe was repeatedly hammering me over the head with the microbiome until I got it.

Once I returned from Germany, I severely restricted my diet, added back one food at a time, and only kept what did not cause a big knot in my stomach (an early sign of future gastritis that I was now able to detect). At the same time, I started a rebiosis program using two different probiotic + prebiotic supplements. Diet soda was gone, as were lactose and gluten, with other grains severely restricted. That combination plus the probiotics did what nothing else could. In 2014, at age sixty-three, I experienced the healthiest year of my life since my twenties. Once I got a nice balance between the probiotics and foods, I was able to use the probiotics only when I detected that knot in my stomach or any other sign of gastric reflux. The probiotics resolved the knots even more effectively than combinations of antacids. The need for antibiotics was drastically cut, as was the number of days I spent either in distress or feeling ill. At the end of 2014 and beginning of 2015, I had my annual checkups with both my family doctor and my ENT specialist. They were both stunned. My family doctor kept mentioning how he tells lots of his patients to lose weight but only a few ever succeed in doing it. Furthermore, the weight loss is usually only temporary. When I told him it was all a result of my Germany observations, plus my microbiome seed-and-feed strategy, he asked for the microbiome papers as he wanted to help other patients.

The visit to the ENT clinician was even more memorable. His entrance began with the acknowledgment that something significant must have happened to me as his office never saw me anymore. It had been more than a year. When I told him that his theory about the gastric reflux combined with my Germany experience and knowledge of the microbiome had shown the way, he was stunned. But he also recounted how he suspected that other patients in his practice were experiencing similar gut-related, inflammation-driven respiratory problems. He could not get around how simple the solution had been. I still need some allergy meds, but the risk of sinusitis is now manageable, where before, it was totally out of control.

This is only one personal example. But given the knowledge that misregulated inflammation is at the heart of most NCDs and that the microbiome educates the immune system in matters that affect control of inflammation, it is likely to match many other cases. There are two take-home messages: (1) Since you have information about the microbiome that I lacked in 1980, don't wait thirty years to do something useful about the 90 percent of you that is microbial. You can work with your health care providers to right the ship when it comes to your microbiome. (2) Changes in diet plus rebiosis are likely to be more effective than either dietary changes or rebiosis alone.

14. Is There a Risk in Trying Rebiosis?

Yes. The reality is that there is some positive risk in almost every aspect of life, including getting out of bed in the morning, driving to work, entering your credit card number on a hotel's Internet website, eating that yogurt, or undergoing FMT. So encountering something with more than zero risk is normal and may not be the critical factor in determining whether to pursue some level of rebiosis. The more useful question is whether the risk that exists is an acceptable risk. Is it acceptable to you? That is all that matters.

One of the more intriguing aspects of our society is how we deal with risk. There are several contributing factors to this, including our own personalities, the levels of actual risk, our perceptions of risk, and who controls the risk (i.e., us versus them). In this chapter, I will discuss risk in general and then as it involves the microbiome. Because the risk is not zero, there is always a possibility that something will happen. That possibility may be ridiculously low, but it is never zero. So a useful discussion of risk begins with the fact that we deal with different levels of risk that are more than zero. How much risk we undertake is based both on our level of tolerance and the likelihood and severity of the event itself. Our approach to risk can also change with age and life status.

Teenagers and young adults are often vested with an air of invulnerability. They are less likely to feel they need various forms of insurance and may be far more likely to ride the newest, most intimidating amusement park ride. Broken bones and falls happen, but healing happens as well. For the teenager's great-grandmother, broken bones and the same fall can be a serious and justified fear. Middle-aged parents of those teens are likely to have a different view of risk and insurance as they view their invested stake in supporting their children's emerging futures.

The big comparison to make is the risk of rebiosis (often low) against the alternative risk of doing nothing. In this case, doing nothing usually means dealing with one or more often-predictable NCDs, including what they mean in terms of shortened life span and reduced quality of life. No action has quite predictable consequences. Still, it is important that any action be recognized as having some risk greater than zero associated with it, and that amount of risk should be accepted before taking a bite of that yogurt, eating a fermented food (that topic is covered in a later chapter), or doing something more comprehensive in terms of rebiosis.

15. Does Rebiosis Work in Animals, Including My Pets?

Yes. Of course much of the initial rebiosis research was performed in mice and rats, which are animals. But what you may not know is that the poultry industry has been using a quite successful rebiosis program for thirty to forty years. You probably have heard a lot about the sad tale of antibiotics in poultry feed, which I discussed in Chapter 9. But there is a much happier and remarkable story about the use of probiotics for decades in poultry. Not only do they work, they have transformed the livelihoods of some poultry farmers and provided a proof of biological principle for we humans who take probiotics.

Earlier in the book I introduced the famed Ukrainian immunologist and father of natural immunity, Elie Metchnikoff. But what I did not include in the prior mention is that Metchnikoff was an early proponent of and personal experimenter with probiotics. He regularly consumed his own probiotic-laden yogurt-like drink, which he was convinced would support a longer life. But if Metchnikoff was a personal pioneer in use of probiotics, it was chickens that first enabled the mass application of probiotics to minimize risk of disease. In 1973 Professor Esko Nurmi and Markus Rantala from the National Veterinary Institute, Helsinki, Finland, published an article in the scientific journal *Nature* demonstrating what came to be known as the Nurmi concept. These scientists were reacting to a devastating outbreak of *Salmonella* in poultry that occurred in 1971. Professor Nurmi was concerned with the fact that the modern agricultural practice was to raise newly hatched chicks apart from their mothers. These age-isolated newly hatched chicks were slow to develop their normal intestinal flora. This made them very susceptible to colonization by bacterial gut pathogens like *Salmonella*.

The comprehensive environmental management of the henhouse meant that you could feed prebiotics and probiotics at will and test out various combinations to find the ones best tailored for specific ages and functions (newly hatched chick, egg-laying adult hen, rapidly growing juvenile broiler, exceptional breeding stock, etc.) to boost health and productivity, ensuring profits for the farmer. The probiotic process used in animal agriculture such as poultry is known as competitive exclusion, where you seed and feed what you want in the chicken's digestive tract and swamp out the capacity of bacterial pathogens to get a foothold.

Following the Nurmi discovery there was a flood of lab research, field trials, and supplement development. By the mid-1990s a number of trials had been reported, and by 2003, there were already extensive reviews of prior work with *Lactobacillus* and *Bifidobacterium* supplementation along with specific prebiotic feeding. One of the driving goals was to reduce the risk for the spread of *Salmonella* both among poultry and to consumers from meat and eggs. But other benefits were also observed, including improved egg production, improved conversion of feed to muscle in broilers, improved overall growth rate, and increased resistance to disease (e.g., reduced mortality among production flocks). This is essentially the profit margin for chicken farmers.

To achieve effective competitive exclusion against pathogenic gut microbes, the probiotic formulations are usually administered to newly hatched chickens or turkeys as soon after hatching as possible. The probiotics can be given in a spray in the hatchery or administered in the drinking water of the chicks. However, the drinking-water route can produce more uneven benefits. Chicks may not drink water until after they've been shipped or even after they've begun eating. For this reason, spraying with probiotics even before the chicks are removed from the incubator/hatchers is a good option. There is interest in injecting probiotics directly into the egg as the embryo is developing, just before hatching. This has had mixed results as some probiotics given to the embryo can interfere with hatching itself. However, more recent efforts provide a glimmer of hope for late-embryo dosing. As with both chicks and children, the timing around hatching and birth are superb windows for action in support of a healthy microbiome.

Competitive exclusion is a bit analogous to a bar scene where someone yells "drinks are on the house," and there is a mad dash toward the bar and bartender. In this case the bacteria that usually cause diseases are not as efficient at grabbing bar stools and they don't get served. Put another way, competitive exclusion is like a promotional contest to log into a new advertising site. The first hundred log-ins win a free Caribbean cruise. In this competition, the probiotic bacteria are on high-speed connections, and the pathogens are using dial-up.

The process has been developed to the point where commercial probiotic products are administered with some containing as many as 200 different bacterial species. A retrospective look at the history of competitive exclusion in poultry concluded that it was highly successful not only for controlling gut pathogens in the birds but also with the added benefit of reducing mortality and enhancing growth. It has even been used in older birds that required antibiotic treatment as a means of restoring the birds' microbiome. This form of rebiosis is particularly relevant to human health.

More than forty years of field experience with competitive exclusion in poultry has proven it to be a worthwhile health strategy. The poultry experience with probiotics has not gone unnoticed. What started as a poultry strategy has not stayed in poultry. Competitive exclusion has been used in pigs to reduce both *E. coli* and *Salmonella* problems in baby pigs. It seems to be an effective replacement for routine administration of antibiotics in feed within the pig industry. Probiotic mixtures have also been useful for calves, particularly those under management stress. What seems to be important is that the probiotic cultures for chickens, pigs, and dairy calves must be derived from the same animal species.

These successes using probiotics to shape the animal microbiome across decades, with massive numbers of animals constituting a significant percentage of the world's animal protein food source, then raise the question, why have we been so slow to formally embrace a parallel strategy within westernized medicine? If a farmer's profit margins and food safety rest on decades of proven strategies of using probiotics, what exactly are we waiting for?

Note that after thirty to forty years of successfully using a probiotic-management approach for poultry, farmers now have the optional approach of feeding microbial metabolites instead of whole live bacteria. The short-chain fatty acid butyrate can be fed to both chickens and pigs to discourage *Salmonella* from gaining a foothold. Of course that chemical, butyrate, still stinks, so it has taken some additional research to ensure that the chickens don't smell as well.

Much as with humans, when your dog or cat has an NCD, the microbiomes of their gut and/or skin are skewed as well. Not surprisingly, probiotics work very well in dogs, and their microbiome management is an increasingly important component of their overall health given the high incidence of immune and gastrointestinal diseases in dogs. Probiotic microbes are now included in pet nutrition, therapy, and standard care.

For example, much like in chickens, oral probiotics are used to treat lower urinary tract infections in pets. One of the same probiotic mixes reviewed in terms of human health (VSL#3) is reported to be helpful in treating inflammatory bowel disease in dogs. Comparison of microbiological, histological, and immunomodulatory parameters in response to treatment with either combination therapy with prednisone and metronidazole or probiotic VSL#3 strains suggested a benefit of probiotic mixes for dogs with idiopathic inflammatory bowel disease. A wealth of US patents relating to probiotic bacteria for dogs have been filed for new product development and protection of intellectual property. This suggests that additional data, presently not in the open scientific literature, exists supporting the benefits of probiotic bacterial isolates in canine health.

In cats, dysbiosis of the gut microbiome has been associated with diarrheal disease with specific major differences in the prevalent bacterial species between healthy versus sick cats. Probiotics are used by numerous different groups within veterinary medicine as part of an integrated strategy to manage both GI tract and renal diseases. The Cornell Feline Health Center includes probiotics as a suggested part of newer treatments for inflammatory bowel disease in cats.

16. What Probiotics Should I Take?

This is the big one. It's the first question I am asked after my lectures and the one I dread the most. Why? First, the answer to that question lies solely between you and your professional health care providers. I can comment on what I see in the research literature and the personal approach I have taken, but that information is not intended as, nor should it be taken as, personal medical advice. What works wonderfully for one person might not work as well for someone else, and I hope the reasons why are now rather obvious.

Second, there are few standards for commercial probiotic products at the present time. It is much like selecting a hotel room in a city you are visiting for the first time. How do you go about that? Would you use online review sites such as Trip Advisor, go only with large hotel chains you trust for standards, or use word-of-mouth recommendations from your well-traveled friends?

Simply to provide you with one example, the approach used in our family has been to (1) first and foremost look at all the scientific evidence presented in a number of research and clinical trial papers suggesting which combinations of microbes are likely to be most useful, (2) use our own microbiome analysis results, and (3) look for any customer reviews about the specific commercial products we have identified as containing probiotic species we want to try. After that we let our own bodies, sometimes aided by relevant biological or microbiological analyses, tell us about the usefulness of the product.

Health Canada generated a list of probiotic strains for which there are sufficient studies to conclude that general benefits can be expected with sufficient doses. These include *Bifidobacterium* (*adolescentis, animalis, bifidum, breve,* and *longum*) and *Lactobacillus* (*acidophilus, casei, fermentum, gasseri, johnsonii, paracasei, plantarum, rhamnosus,* and *salivarius*). This does not mean that consuming other microbial strains has no benefit. It is only that these strains have substantial evidence as per benefits.

The bottom line is that even with some uncertainties, more information to come out, and risks greater than zero, there is little reason to simply live with NCDs and treat the end-process symptoms with heavy-duty pharmaceuticals while never addressing the root of the problem. It is time to consider treating yourself to a superorganism makeover.

13

TO BE A MICROBIOME WHISPERER

Once you understand how your microbiome works, you can gain a sense of mastery over the entire co-partnership. That is where really effective self-care emerges. As I learned from my own personal path with NCDs and my microbiome, this can be really self-empowering.

Your microbiome is your co-partner and there are some deep secrets and behavioral quirks that you need to know about it. Since you are going through life together, it could be time to set some rules and boundaries when it comes to the shared living space that is your body. I don't know if your body is your temple, but I do know it is the house where your microbiome lives. It's time to appoint yourself head of the homeowners association (HOA) for your body. Your microbiome can't go off throwing wild teenage parties and trashing the house while your attention is elsewhere, dealing with life's stresses. To ensure helpful behavior, you need to train your microbiome just like you would train your kids or your pets, all the while becoming its caretaker and protector and enjoying life together. Training your microbiome is a little like becoming an effective parent or an effective dog trainer. You need to have a good handle on the trainee. You need to understand certain unhealthy tendencies and behaviors, recognize when they arise, and have a plan to minimize any damage and turn those behaviors into something productive. Earlier in the book, I emphasized your mighty majority microbes. But for self-care, now is the time for your minority mammalian self to take charge. You can and should be respectful of your microbes but, nevertheless, be fully in charge of your own body even if the microbes outnumber you. You want to understand microbe behavior much like Cesar Millan understands dog behavior. There are horse whisperers and dog whisperers—and there will be microbiome whisperers.

Sometimes microbes operate not just independently but as a pack, and they can easily shift into a pack mentality. Quite often that is not a good thing for your body. That is where you lose control just the same as if your pet dog were enticed to join a neighborhood pack of dogs intent on bringing down a deer. But just as Cesar Millan knows dogs, your knowing about your microbiome's pack tendency allows you to become a master trainer of your microbiome. To really get the upper hand, bringing the microbial part of you in line with your health goals, it is useful first to think like a microbe and then to set some goals. What does the social life of your microbes look like? How do they live, socialize, and protect themselves? What kind of microbes do you really want to cohabit with?

Your Enterotype and Your Microbial Diversity

Different people from different geographic areas with long experiences with different diets have different lead bacteria in their guts. They have different enterotypes. It is clear that people from different ethnic groups, geographic regions, and even lifestyles will host different populations in their microbiomes. This is not shocking since the microbiome, or second genome, is a product of ancestry and environment (the latter also including dietary patterns). For this reason, what constitutes an ideal, normal gut microbiome in Kunming, China, is not necessarily identical to one you would find in healthy adults in Boulder, Colorado; Maracaibo, Venezuela; or Adelaide, Australia. Different ethnic makeup, different environments, and different diets result in overlapping yet different profiles for the microbiome.

When you consider working on your gut microbiome, you are likely to find that some changes are easy to make and sustain while others just don't seem to take hold. Researchers believe that is the difference of working within your predominant pattern of gut microbes (i.e., your own enterotype) versus working across enterotypes. You may not know your enterotype at the moment, but there are lots of companies offering microbiome analysis services that can tell you. You don't have to know your enterotype to change your microbiome. It is simply something to be aware of as you pursue self-care.

However, there is a second useful measure of microbiome health, and that is its richness or diversity. It is not enough to get your lead gut bacteria in line with your diet and ancestry. You need to ensure that your microbiome has the diversity needed and the right microbial players in place to perform all the necessary co-partner functions for your body. This goes right to the core of the rare microbe in your gut performing a critical function being the microbe you most want to nurture and protect. That is, in effect, knowing your weakest link. There are plenty of the most prevalent bacteria in your gut around, but always having enough of the really rare, functionally important bacteria can be the difference between health and disease. In training your microbiome, these two concepts, knowing your lead bacteria and keeping your microbiome diverse, need to coexist.

Evidence suggests that microbial diversity is generally beneficial and should be one of our training targets. In other words, it is helpful for us to have enough different species and strains of bacteria and archaea to provide all the metabolic pathways, neuroactive chemicals, and immune signals that our body needs to be in balance and to function well. If we are not sufficiently diverse in our microbiome, that is usually a warning sign of an impending or existing health problem. There are two important pieces of evidence. First, many different categories of diseases (e.g., allergic, autoimmune, inflammatory, metabolic) are associated with microbiome profiles in which bacterial species are missing or the numbers of key bacteria are too few to perform vital tasks. Although high diversity of microbes does not always guarantee good health, it is often a health risk to have a restricted diversity of microbial species compared with healthy controls.

Second, an international team of researchers recently analyzed the fecal, oral, and skin bacterial microbiomes of a remote group of Yanomami Amerindian village people living in the Alto Orinoco region of the state of Amazonas in Venezuela. These people had no known prior contact with people of European descent. These indigenous people of the Amazon jungle spanning Venezuela and Brazil had the highest diversity of bacteria ever found in a group of humans. Genetic analyses also suggested that they had a broader array of genes at their disposal for biological functions. Are the indigenous people disease-free? Certainly not. They die, but it is usually of infectious diseases. While some of the infectious-disease mortality is due to local pathogens, a significant portion of it is also caused by newly encountered pathogens introduced via contact with outsiders. You might say their mortality profiles look like a European population of many centuries ago.

What they don't have is important. It is our current global epidemic of NCDs. In fact, hypertension and obesity were nonexistent among the Yanomami sampled in recent decades. However, a 2014 study compared obesity rates among jungle Yanomami with those now living in two villages with a westernized lifestyle (called transculturation). As had been seen in earlier studies, there was no obesity among jungle adult Yanomami. In contrast, there were high rates of obesity (44 percent and 89 percent) in adults from two different villages with different degrees of nutritional transitions to a westernized lifestyle. The take-home message is that there are lifestyles where the NCD epidemic does not exist, and those seem to feature the combination of a nonwesternized diet supporting a more diverse microbiome. The goal will be to identify the changes that get us back closer to living in nature while still reaping the benefits of civilization's progress. There is some middle ground where the NCD epidemic is defeated.

Probiotics have been suggested as tools to shift our physiology from a pro-tumor-growth state to an anti-tumor environment, particularly when it comes to gastrointestinal cancers. In a human study demonstrating that probiotics can shift the gut microbiomes of colon cancer patients, researchers in Shanghai, China, showed that tissue from cancer patients had restricted microbial diversity compared with healthy controls. Additionally, the cancerous tissue was dominated by bacteria from the genus *Fusobacterium*. Because heavier loads of these specific bacteria in this type of cancer are associated with poor immune response against the tumor and shorter patient survival, it has been suggested that *Fusobacterium* load is a helpful indicator of patient prognosis. Taking a probiotic both reduced the presence of *Fusobacterium* and increased the density and diversity of gut microbes overall. Determining whether probiotics can actually extend the life of colorectal cancer patients will take more study. However, these results suggest that a mix of microbes that seem to promote tumor growth and prevent immune attack can be changed using probiotics.

Drawing upon the human garden analogy, you will be training your microbiome garden, in general, to be prolific in terms of numbers of each vegetable and more diverse in terms of types of different vegetables (microbes).

Also, pay attention to your soil and climate type as you work on the microbial garden (that would mean playing to your enterotype as you work toward a healthier balance of microbes). Shifting within the enterotype may be easier than shifting between enterotypes just because they have been engrained in your ancestry for a very long time. The good news is that healthy microbiomes exist within all major enterotypes examined to date.

A good example illustrating why it may help to work within an enterotype comes from a recent analysis of gut microbiota from 303 school-age children living in rural and urban areas from five different countries in Asia. The microbial profiles of the children fell into two major enterotypes, those dominated by *Prevotella* bacteria (P-type bacteria) or those dominated by *Bifidobacterium* or *Bacteroides* bacteria (B-type bacteria). Subtypes were found within the two major enterotypes. A majority of the children in China, Japan, and Taiwan had the B-type lead bacteria enterotype, while a majority of children in Indonesia and the Khon Kaen region in Thailand had the P-type lead bacteria enterotype. Notably, each major enterotype included children who had a healthy body mass index as well as those who were obese. So in training your microbiome, you can move from an unhealthy state to a healthier state yet stay within the same major enterotype (lead bacteria group).

Beyond the lead bacteria there were other country-of-origin and rural-versus-urban differences. For example, a bacterium known as *Dialister invisus* was detected from 67 percent of children in Japan but only 18 percent of children from cities in other countries. Some differences were found that might relate to the type of rice that was eaten and its resistant starch content (which varies between the rice in Japan and Indonesian rice). Also, the research found a distinct rural-versus-urban difference in the predominant gut microbiome of children within Thailand. Most children in rural Thailand had the P-type profile while most children in Bangkok had the B-type profile. But in a separate study of Bangkok residents focusing on vegetarians, their gut microbiome looked more like that of people from rural areas of the country (who also ate more vegetables than their counterparts in Bangkok). This study contributes to the idea that cities in general are not that beneficial for our microbiome except among the few who go out of the way to lead healthier lifestyles (such as creating a personal farm-type lifestyle in the city).

In a recent study, two different groups of healthy children in geographically distinct areas of Thailand were analyzed for their gut microbe composition. In the northeast part of the country people eat different meats, a wide variety of carb types (including fermented rice), and a diversity of fruits and vegetables. In contrast, in the central region of Thailand people eat more rice, breakfast cereals, and cow's milk in their diets. Distinct differences in the microbiomes were associated with these different region-based diets. One of the differences between children was higher representations of *Lactobacillus* and *Bacteroides fragilis* in the northeast. It is important to keep in mind that the span of these dietary differences would involve not just the children in the samples but their parents as well.

Other comparisons of the gut microbes among children on two continents and in four different regions showed that the region can be as or more important than the continent in affecting the features of the microbiome. In this instance, the exact cause (i.e., diet, latitude, other factors) of the regional differences is not known.

It is not just gut microbes that are affected by geography, climate, and diet. A climate and geography comparison was made of the microbes found in saliva among adults from Alaska, Germany, and Africa. The result was that Alaskans and Germans were more similar to each other than either area was to the Africans. However, there were core groups of microbes that were always seen together regardless of the sample locations.

All this research shows that, ultimately, the goal should be to match specific dietary adjustments and rebiosis strategies with your enterotype whenever possible. Have them meet you where you are now. Don't try to find some perfect magic elixir that works for everyone on earth. You would be working against centuries of superorganism programming. Find the mix that works for you personally. If it is time for more personalized medicine, it is also time for more personalized self-care. Make any self-care recommendation prove itself by the results in your body. You will feel them.

Influencing Groups of Microbes

We can use the group behavioral tendencies of microbes to our benefit. Like people, and cows and fish and insects, microbes communicate with one another and can act both individually and as groups. This communication goes on both inside and outside of your body. You want your microbes associating with the right crowd and not engaging in any gang-like behavior that would damage you. One of the processes through which the microbes communicate and act in union is called quorum sensing. The name literally refers to the fact that our microbes can detect when other microbes are around, as well as the types and numbers of those other microbes. They can then decide if they will participate in group microbe projects. Some of the projects may be in your best interest, some projects not. That is where you as a properly prepared trainer of the microbiome need to step in.

Quorum sensing is a process of communication among bacteria and archaea that influences group behavior. It allows microbes to detect the density of populations in a given environment—your body, for instance—and to detect and signal other environmental changes. Because they are in constant communication about the surrounding conditions, they can coordinate responses that simulate that of a whole intact organism. Among the changes that bacteria undergo is to alter their metabolism based on the available nutrients, avoid the accumulation of toxic chemicals, and protect themselves against other microbes. Pathogens will use quorum sensing to defend against the human immune response and to increase their capacity to infect rapidly (often called virulence factors). There are literally chemical circuits that can become activated at the same time in thousands of microbes that promote group changes and action. Different types of bacteria (e.g., gram-positive versus gram-negative bacteria) utilize different forms of quorum sensing systems.

In one type of quorum sensing strategy, proteins called autoinducers are produced by the bacteria, aiding communication with other microbes. It is a process that can result in changes made as a herd. In some cases these changes can be beneficial for human health and in other cases can present a greater chance of disease. By understanding how and when quorum sensing works, it is possible to manipulate those signals to help maintain a healthy microbiome and reduce the risk of dysbiosis-induced disease.

One of the examples where a pathogen uses quorum sensing to increase virulence is found in the skin bacteria *Staphylococcus aureus*. Normally, the bacteria results in only minor skin infections if the skin barrier is broken, allowing its entry. However, under different circumstances, it can lead to serious, life-threatening infections. *S. aureus* is a major cause of infection within hospitals. This bacterium can turn pathogenic when different combinations of the quorum sensing genes are activated and the virulence of *S. aureus* increases significantly.

Comparisons have been made among *Lactobacillus* species, where certain species are rigid in their physiological responses and others can seemingly adapt to different niches. In a study from the Netherlands investigators discovered that the *Lactobacillus* species *L. plantarum*, which is found not only in the human gut but also in fermented foods and plants, contains more quorum sensing genes and related components than do other *Lactobacillus* species with highly restricted niches (e.g., *L. johnsonii*).

The quorum sensing molecule produced by bacteria may be recognized by human mammalian receptors as well. At least one study suggests that the mammalian part of humans may be able to listen in to bacterial chatter transmitted by quorum sensing molecules. Much like keeping track of your tweens' and teens' activities on social media such as Twitter, Instagram, and Facebook, it is good to stay on top of quorum sensing chatter within your microbiome.

While quorum sensing is a natural strategy for cooperation among microbes, specific knowledge of the pathways involved in quorum sensing allows a new opportunity for microbiome management. There are three main ways in which quorum sensing can be used to improve health and reduce the risk of disease: (1) provide indicators of impending changes in our microbiome, (2) provide potential drug targets to block impending or ongoing infections, and (3) provide new opportunities to adjust the composition and status of the microbiome in different tissues.

In the first situation, quorum sensing signals can give us a measure indicating that potential harmful changes are under way in our microbiome. The signals are what are called biomarkers and could be used to help distinguish between harmful and healthful environmental exposures (e.g., potentially harmful exposures to environmental toxins or useful ingestion of certain probiotics) as well as determine the effectiveness of medical therapies. Depending on the microbes involved, certain environmental exposures could change quorum sensing in a way that would present us with an impending health risk.

One of these is a tendency of certain pathogens to gang together, change their own physiology, and form what are called biofilms. Biofilms are very difficult for us to attack immunologically or through the use of antibiotics. But there may be a solution to biofilms found in our increasing understanding of quorum sensing. For example, some of our gut bacteria have the capacity to block biofilm formation by gut pathogens. Among the products they produce is an antibiofilm enzyme called acylase. We can use our microbes' own strategies to police destructive gangs among pathogens and control unruly microbes within us.

Measuring quorum sensing signals gives us a powerful tool. Changes in these signals could be an early warning signal that could help prevent disease. Other desired quorum sensing signals could help us to determine if a given probiotic is producing a desired outcome in the gut, skin, airways, or urogenital tract.

Here is a real-life example. By using altered quorum sensing signals, it should be possible to interfere with the conversion of pathogenic bacteria into dangerous infective agents. If these bacteria can't act as a group and bind tightly to epithelial cells lining our tissues, the bacteria can't form biofilms to thwart immune attack. With such interference, pathogenic bacteria lose their tools and advantages for producing disease.

Here is an example of how gut bacteria can use quorum sensing to block a pathogen. In an intriguing study involving both Bangladeshi children and germ-free mice, investigators from three continents found that a specific commensal bacterium in the guts of normal healthy children in Bangladesh (*Ruminococcus obeum*) can use intermicrobial communication to reduce the ability of the cholera-causing bacterium *Vibrio cholerae* to colonize and cause disease in the human gut. The commensal bacteria produce a quorum sensing effect on the *Vibrio cholerae* that changes its gene expression and restricts its ability to become established in the gut and cause disease.

I suspect these types of natural and deliberately manipulated, microbiome-centered strategies will be used to reduce the risk of some infectious diseases in the future.

Microbe Memory and Self-Defense

Not only humans, animals, and plants need to worry about attack from viruses. Bacteria and archaea inhabiting our body can be invaded by viruses as well. While we rely on our multicellular immune system to protect us from viral and pathogenic bacteria infections, our own microbes don't have the luxury of lymphocytes and macrophages and all the many different kinds of immune cells that we have. Are they totally defenseless? It turns out they have a rather ingenious plan.

Bacteria and archaea have their own equivalent of an immune system, only they don't have any army of specialized immune cells to recruit and send forward into the attack. Instead, they mobilize different types of enzymes to literally cut viruses to shreds. Where we use cells to attack our enemies, they use enzymes. Their system is called CRISPR.

CRISPR stands for clustered regularly interspaced short palindromic repeat, which is a type of immune system for prokaryotes (single-celled microorganisms without a nucleus, such as bacteria and archaea). Like the human immune system, CRISPRs can recall having seen an outside threat before. This is called immunological memory. With this memory the second exposure to the same challenge (such as infection with a virus) allows the immune response to be more specific against the pathogen, move faster, and utilize more resources. In this case, the bacteria want to be protected from viruses (known as bacteriophages) and other mobile pieces of DNA that could compromise the bacteria's integrity and/or subvert their functions.

In some ways the bacterial CRISPRs' attack on viruses looks a little like the metal-munching Sentinels mobilized against Zion in the *Matrix* movie trilogy. The enzymes rip through viral DNA, destroying viruses and helping the bacteria and archaea to maintain their integrity. But the story behind these bacterial enzymes is proving to be much more than it originally seemed.

If recently revealed secrets of the microbiome have already spawned a revolution both in safety evaluation and health care, then there is even more to be gleaned from our microbial partners. The striking discovery that bacteria have their own type of immune system has paved the way for new human and animal therapies as well as plant science technologies. Because bacteria are highly susceptible to attack by viruses, they need a way to protect their own integrity. For this purpose, they have developed a unique genetic-based strategy for protecting themselves from these attacks, as discovered by Jennifer Doudna at the University of California, Berkeley, and Emmanuelle Charpentier at the Helmholtz Centre for Infection Research and described by Carl Zimmer in *Quanta Magazine*.

This protection involves the bacterial capacity to capture pieces of DNA from an invading virus, store them in specific places within their own bacterial genome, convert the viral DNA copy into copies of RNA, and then mobilize the RNA pieces along with specific DNA-digesting enzymes to attack the DNA of the same invading virus. The RNA sequences exquisitely match the viral DNA such that the enzymes only destroy the DNA of interest. For the bacteria, this is a specific defense with little energy wasted and few adverse side effects.

The entire process relies on two series of gene sequences. The first are the already described CRISPRs. Next to these are genes for producing the DNA-cutting enzymes called Cas, which stands for CRISPR-associated genes. These enzymes troll the bacterium, carrying the RNA copy made from viral DNA as a landing pad. Once this landing pad latches onto the matching viral DNA, the enzyme goes to work cutting the DNA into pieces and destroying the viral genome. It is the Cas enzyme seeking a precise match for the RNA that brings specificity into the attack. CRISPR and Cas make an effective team. A particular Cas, Cas9, appears to be very important in giving the whole immune-like defense system its capacity of memory.

Notably, this bacterial immune defense is specific against the single invading virus and does not destroy other DNA. Because of this specificity and the fact that the bacteria use prior exposure to the virus to their own advantage, the bacterial defense represents a type of adaptive immune response. In fact, there appears to be a type of bacterial vaccination event that occurs when the viruses that invade bacteria (also called bacteriophages) have defective phages. Exposure of bacteria to these defective phages can lead to CRISPR sequences being set up but give the bacteria time to get ready for a real intact virus attack. Also, there is selection against what would be considered an autoimmune type of response (where the sequences and enzymes overlap with the host bacterial genome).

The CRISPR-Cas9 system of our bacteria and archaea has many overlaps with our own immune system. Besides the aspect of memory of prior attacks, bacteria are able to select against their own DNA and in favor of foreign DNA to place in the CRISPR area of the chromosome. The bacterial CRISPR-Cas9 system protects the bacteria's genes from attacks by invading foreign genetic material. The system also overlaps with the way our own mammalian immune system interacts with the external environment, including our microbiome. The human immune system is known not only to protect against pathogenic invasion of the host but also to create a steady state interaction with the environment and proper function within tissues. In many ways the CRISPR-Cas9 system of bacteria mimics this physiological function as well. Researchers have found that CRISPR can be present even in the absence of viruses. The question is, what is it doing in those situations? One idea is that it allows bacteria to respond to environmental changes by changing their own cell envelope physiology.

CRISPR-Cas9 will have biotechnological applications well beyond the microbes and will affect future medicine and therapies. In effect, we are learning from our microbes how to better battle disease. But for now it is helpful that this gene-based strategy of our microbes allows them to function as if they had their own type of immune system, which can sample their environment, interact with it, and protect them from environmental attack.

These remarkable developments in our understanding of the technical details of microbe behavior are what enable us to personally take charge of the microbes within us. But another line of research has perhaps even more amazing implications for our daily sense of well-being. Let us turn to the effect of the microbiome on our psychology.

14

YOUR BRAIN ON MICROBES

n March 2013, the Public Policy Polling group surveyed 1,247 American voters regarding their various beliefs. Among the question topics were ideas about who was really in charge of controlling our lives. The result indicated 28 percent believed there is a secret, elite authoritarian group conspiring to rule the world; 15 percent believed that the government adds mind-control technology to media broadcasts; and 4 percent believed there are shape-shifting, interplanetary aliens running the world. Of course, the emphasis was on who runs our show from the outside; what was not asked in the poll is who runs our show from the inside. Put another way, who exactly is running your superorganism? I wonder what percentage of those 1,247 voters suspect that microbes are affecting virtually all of their decisions.

Several researchers, including John Cryan at University College Cork and Carlo Maley at the University of California, San Francisco, have referred to the microbiome as our puppet masters. John Cryan's colleagues at University College Cork point out that there are at least five puppet master strings or routes of communication between the gut microbes and brain: (1) immune signaling that also includes the hypothalamus-pituitary-adrenal (HPA) axis, (2) activation of the vagus nerve, (3) spinal pathways, (4) direct production of neurotransmitters by bacteria, and (5) microbial production of short-chain fatty acids. These routes provide not simply the transfer of information but also real change in our brain physiology and function.

We have known for some time that control of behavior can be circumvented by some parasites and pathogens. They can take control in a selfish way. An example most people know about is the case of the rabies virus, which causes its victims, including humans, to modify their behavior, become aggressive, and even bite. These actions help to spread the virus between animals and people through saliva. It is thought that the aggression is brought on by a severe reduction in serotonin levels. It is why you don't take chances with animal bites, and why our diagnostic labs at Cornell do a great deal of testing for rabies in bats, dogs, foxes, skunks, and other warm-blooded animals. Stephen King made good narrative use of this virus in his rabid-dog novel-turned-movie titled *Cujo*—for older generations the Disney dog film *Old Yeller* covers the ground in a more tender way.

If pathogenic microbes can control our behavior, so can our microbial friends, and there are many more of them. The challenge is for us to determine exactly what is ultimately malicious and what is benevolent when it comes to a microbe's intent.

The gut, with 70 percent of our immune system and thousands of microbial species, is critical to our health in a world filled with friends and foes. Following that fundamental requirement, the gut-brain axis is the next most significant connection since it colors how we feel about ourselves within our world. There is considerable stock placed in personal actions and responsibilities, as that is a key part of what holds societies together. However, the "personal" part of our personal responsibility involves a cast of thousands of different species.

The gut itself so affects neurological function and the brain that it has been called the "second brain." Yet buried within the gut are our gut microbes, and in many ways, they are the puppet masters hiding behind the throne. Our microbes have been working with and on our ancestors for centuries. In fact, they have been working on our ancestors using epigenetic gene switches that very likely transferred down through generations of our predecessors, as described in Part One. The microbes developed strategies to mold us into the ideal co-partners, taking a very long view of this human behavioral project.

Food Cravings

If a roof over their heads and three square meals are basic needs, then our microbes would prefer to stay home and simply send you out foraging with a type of grocery list. And they know how to do precisely that. Dark chocolate is rich in a variety of chemicals, including polyphenols. These are part of the same group of chemicals found in various fruits, red wine, and grapes and have been known for some time to play a role in preventing NCDs. Our gut microbes have to act on these chemicals for them to be really useful to us. Recently, researchers have pointed out that our gut microbes make these useful chemicals for us, and in turn the polyphenols affect the state of our gut microbes since they need these chemicals, too.

Our gut bacteria make the dark chocolate healthier for you as long as your calorie intake remains in check. There is evidence that people who crave chocolate have different microbes in their gut and different chocolate metabolites in their urine. It seems clear that your profile of gut microbes, your desire for certain foods, and the metabolites that you produce from that food all align. Exactly which signals from the gut microbes affect the relationship to cravings has yet to be determined among myriads of possibilities. Since the microbial by-products can drive our contrasting feelings of pleasure, euphoria, depression, anxiety, discomfort, and pain, that is the place to look. The take-home message from this area of work is that by changing your microbes, you have a better chance to change your eating behavior.

The various bacteria in your gut are not simply innocent bystanders hoping you might accidentally choose to feed them over other microbes. They know how to biochemically influence you to choose their preferred food over others. A battle of signals rages inside you that eventually translates into a menu you only think you created. Instead, it reflects the balance of power among your microbes. There are three major phyla of gut bacteria: Actinobacteria (A), Bacteroidetes (B), and Firmicutes (F). It turns out that the ratio of B to F bacteria is much higher in lean individuals than in obese individuals. In fact, a high F-to-B ratio is pro-inflammatory and considered a biomarker for obesity. In general, a high-fiber diet is preferred by and facilitates the growth of B, and a high-fat diet is preferred by and facilitates the growth of F. Studies of children in different regions of the world indicate that those who eat a high-fiber, low-fat diet have a higher B-to-F ratio. But the phyla include many different bacteria, all with their own food preferences and needs. At a deeper level of comparison, gut bacteria within the three different phyla also have their own food preferences.

Preferred food sources differ as well depending upon what the bacteria can use as an energy source. For example, *Prevotella* bacteria, a genus within the B phylum, want to eat carbs, while *Bifidobacterium* bacteria, a genus in the A phylum, crave dietary fiber. Other minor microbial players in our gut (e.g., *Akkermansia muciniphila* and *Roseburia* species) have their own preferences and are weighing in as well. It is a balance-of-power issue. If you have eaten something for generations in your family, your microbes are likely to reflect that. You are synced up with your gut microbes' long-standing energy sources. On the other hand, if you have a certain mix of microbes currently in your gut, they will be working hard to see that you eat precisely what they want.

Chemicals to Control Our Brain

Our gut microbes control much of our neurological and brain function because they produce a wide variety of neurotransmitters and neuromodulators in addition to affecting the production of those same neuro-active substances by our mammalian cells. These brain-affecting chemicals can reach the brain via either the enteric (or gut) nervous system or the portal circulation (a vein that runs from the gut to the liver). An ever-increasing list of neuroactive metabolites of gut microbes have been reported. These include:

> serotonin (made by some *Enterococcus* species)
> dopamine (a product of some *Bacillus* bacteria)
> gamma-aminobutyric acid (GABA) (produced by some *Lactobacillus* and *Bifidobacterium*)
> acetylcholine (produced by some *Lactobacillus*)
> histamine (produced by some *Lactobacillus*)
> norepinephrine (produced by some yeasts, *Escherichia*, and additional bacteria)

Serotonin is a neurotransmitter that regulates sleep, mood, and appetite and also affects certain cognitive functions such as learning and memory as well as cardiovascular function, bowel motility, ejaculatory latency, and bladder control. As a result regulation of serotonin is very important for our health, mood, and well-being—and the historic success of the drug Prozac and other selective serotonin reuptake inhibitors (SSRIs) illustrates the importance of appropriate balance of neurochemicals. What we had not realized is the extent to which our microbes control the balance of our neurochemicals. Much of the body's serotonin is produced by specialized cells in the colon. It turns out that gut bacteria can control the gut's production of serotonin via the metabolites that specific spore-forming gut bacteria produce. You might say that in controlling such central aspects of our core being, the microbes have us right where they want us. However, if we don't like their influence at any given point, we now have the capacity to evict them and install new co-partners.

Accumulating evidence indicates that things like mood, anxiety, and depression are affected by microbes and their specific activities. Besides directly producing hormones and neurotransmitters, other metabolites of our gut microbes can epigenetically program our neurological system for behavioral characteristics. This programming can happen early in life and affect the rest of our lives. A team of University College Cork researchers recently showed precisely that. Germ-free animals lacking normal gut microbes have specifically altered gene expression in the amygdala that is connected to neurobehavior.

Germ-free mice lack the desire and/or capabilities of social cognition. They are antisocial. As John Cryan and his colleagues have hypothesized, social interactions, including a social collective mind, may have evolved for the primary purpose of permitting the exchange of microbes between individuals. Many authors have discussed what is called the collective consciousness: the hive mentality, shared values, and social mind of communities of people. However, it is clear that any future discussion of how collective consciousness and the unconscious operate will need to include our microbes.

This provides a new view on what can happen when infants and children receive broad-spectrum antibiotic treatment, thereby depleting their microbiome. There are both early postnatal brain function effects as well as later effects seen during critical windows of childhood development, all of which are controlled by microbiome status. John Cryan and his colleagues performed that experiment in mice, looking at behavioral effects. They found that antibiotic-induced alteration of the microbiome after weaning (during later infancy) led to changes in the adult microbiome profile and also resulted in cognitive deficits. In fact, besides their later-life cognitive deficits, germ-free mice have eerily similar social interaction profiles to those of autistic children. These findings of the neurological importance of the early-life microbiome suggest that any procedure causing a depletion of a baby's or infant's microbiome could have a lifelong, adverse impact on brain function.

Imbalance in our microbes can also injure the brain via inflammation. Our microbes are a source of both pro-inflammatory and anti-inflammatory signals. But when the balance is shifted inappropriately, as with overgrowth of some bacteria in the gut, the ramifications can be dramatic. Pro-inflammatory signals can increase gut permeability as well as the levels of both systemic and central-nervous-system inflammation. The resident macrophages in the brain, microglia cells, are primary responders to many of these bacterial signals. When inappropriately activated, these cells can promote neurodegeneration and destruction. Depression, mental illness, and neurodegeneration are very difficult to correct until and unless the gut microbial signaling that created the inappropriate neuroinflammation is reversed. Much of the effort to date has been to focus on the site of the inflammatory damage (e.g., brain, gut lining) rather than on the source of the problem, imbalance in the gut microbes. There is encouraging human evidence that shifting the balance of gut bacteria using probiotics can reduce systemic inflammation.

We need to rethink mental health from the viewpoint of what is best for the superorganism.

As this is a self-care chapter, I will discuss how you can use the idea of master-controller microbes to improve your health. What was the bestselling prescription drug in the United States in 2014? According to WebMD it was Abilify at $7.2 billion in sales between July 2013 and June 2014. Abilify, generically known as aripiprazole, is an antipsychotic drug that works by changing the chemistry of the brain, primarily by dampening the signal produced by dopamine receptors. It is used to treat depression, schizophrenia, bipolar disorder, and behavioral issues in children. One of the side effects is elevated risk of suicide in the young. But in the era of the microbiome, now that we are armed with the knowledge that our gut microbes are the master controllers of our brain, heavy-duty prescription drugs are no longer the only option. Change your microbes; change your life.

Our brain is similar to the immune system in one important way: They both need a balanced, healthy microbiome to develop and function well.

The prevalence and toll of neurological diseases and disorders is stagger-ing. In a recent study of twenty countries, the incidence of neurological deaths has risen significantly between 1990 and 2010 compared with deaths due to cancer or circulatory disease. For example, cancer deaths in the United States fell by double digits over the twenty-year span, while neurological deaths rose by double digits. As the researchers have pointed out, the specific increase in neurological deaths over those of other cate-gories may be due to differences in the effectiveness of treatments. That would be all the more reason to examine the value of manipulating the microbiome to protect the brain and neurological system better.

The high percentage of deaths from neurological disorders is partic-ularly evident in the United States. It is presently estimated that one in three seniors in the US dies of Alzheimer's or another dementia. How-ever, death is not the only meaningful measure when it comes to the human toll from the neurological part of our broader NCD epidemic. Perhaps an even more alarming concern and related indicator of our problem is the recent explosion in mental illness and behavioral condi-tions in children. Based on recent estimates, one in forty-five children now has autism spectrum disorder, and up to 20 percent have a mental illness for which care costs $247 billion each year. Given that the upcom-ing generation is challenged with serious NCD issues for which many need continual care, and that the aging baby boomer population is de-veloping their own set of NCD issues for which they will need continual care, exactly who will be left as the caretakers when we reach the tipping point? Continuing the path of ineffective, patchwork solutions for this NCD epidemic is no longer an option.

Rebiosis on the Brain

If microbes affect our behavior and cognition, and we can reshape our microbiome and its metabolism using probiotic and prebiotic supplementation or FMT treatments (by working in concert with health professionals), then we have the potential to change and improve our cognition and behavior significantly. For anyone facing these diseases and conditions, either personally or within their families, this is a sea change in terms of both hope and opportunity. If you don't like your current mental state, there are emerging strategies for shifting the balance of functions in the brain that are alternatives to lifelong, side-effect-producing, heavy-duty medications.

There is research supporting this type of new approach, although it is important to note that not just the species of probiotic bacteria matters but also the specific strain since genes, metabolites, and physiological effects can differ between strains. One useful, newly emerging probiotic is *Bifidobacterium longum* strain 1714, which has an antianxiety action in mice. Additionally, a recent human trial involving probiotics and college undergraduate students found that daily supplementation with the probiotic bacteria *Bifidobacterium bifidum* R0071 resulted in a higher percentage of healthy days among academically stressed students. Finally, a psychiatric study in humans concluded that consumption of fermented foods containing probiotics was associated with reduced social anxiety. A recent study compared the mental health and HPA effects among workers in the petrochemical industry who for six weeks ingested yogurt with probiotic bacteria, an encapsulated mix of several probiotic bacteria, or standard yogurt as a control. Both types of probiotic supplements significantly improved both general and mental health among those workers who consumed them compared with controls.

There is evidence that supplementing with certain prebiotics can boost resiliency to stress by acting through the gut microbiota–brain axis. In a stressed-mouse model, a team of researchers at Ohio State University looked at the impact of dietary supplementation with two different, slightly modified lactose sugars that are normally found in human milk. The two types of human milk oligosaccharides are indigestible by our mammalian cells, pass through the small intestine unabsorbed, and then get used by microbes in our colon to make a variety of needed metabolites. These specific specialized sugars from breast milk are known to support the growth of friendly co-partner bacteria such as *Bifidobacterium longum*. In this study, male mice were given either of the two prebiotics for two weeks prior to a social-disruptor stressor where the younger male mice were confronted with a "foreign," aggressive male mouse. The prebiotic supplementation led to greater maturation level of brain neurons, a stabilized gut microbiome after the social stress exposure, and a lack of anxiety in poststress behavioral testing. These results support the value of nurturing brain-friendly gut microbes via supplementation with useful prebiotics. They also reaffirm the value of breast-feeding whenever possible for neurological benefit.

A study in humans of two prebiotics yielded encouraging results similar to what has been found with rodents. Work-related stress can elevate cortisol levels and also reduce work effectiveness because anxiety makes attention and focus difficult. When researchers compared a prebiotic galacto-oligosaccharide preparation versus a second prebiotic and controls, they found that people taking the galacto-oligosaccharide preparation had both a reduction in neuroendocrine stress response and anxiety-related changes in attention and focus. There is the real possibility that prebiotic intake may provide a useful route to reducing anxiety and improving work performance by disconnecting us from constant fight-or-flight responses.

Prebiotics, like oligosaccharides found in breast milk, can be useful for infants, with manufacturers increasingly adding them to some infant formulas. These prebiotics include galacto-oligosaccharide, fructo-oligosaccharide, polydextrose, and various combinations of these. Positive effects have been reported with the supplemented formulas. For example, formula supplemented with galacto-oligosaccharides enhanced the growth of *Bifidobacterium* and *Lactobacillus* bacteria, inhibited the growth of *Clostridium*, and reduced the prevalence of colic.

These prebiotics can also lead to an increased production of a variety of neurotransmitters and neuromodulators working between the gut microbiota and the hippocampus.

Putting all of this together, there is real promise for probiotics and prebiotics to increase our resiliency throughout our physiological systems, including the brain and neurological system. They can produce combined effects across the immune, neurological, and endocrine systems and help buffer us against the ravages of stress regardless of the type and origin. We perform better in terms of brainpower, focus, and memory, and we are less likely to see major HPA fluctuations in response to the stress that hammers our immune defenses and opens us to more disease.

15

WILL YOU DO NO HARM?

Keeping humans safe, alive, functioning, in good health, and enjoying life is something we have been doing for ourselves and others for millennia—except that the "human" that families, communities, and civilizations have been trying to keep safe we now know to be a different animal, a superorganism. Keeping a complex, multispecies superorganism safe is a completely different ball game than just protecting a single-species mammal. What do you think is safe for the thousands of species in your body, for your microbial co-partners in life? In almost every case, the data that have been and are currently being collected by government regulatory agencies give us no clue. They may tell us what is safe for mammals but ignore the risk to the microbiome and the signs of a damaged microbiome. They have not been conducted in a way to ensure protection of your microbiome. Due diligence is not being done. No longer can our government, nor the food industry, nor pharmaceutical companies, nor the medical profession plead ignorance. And neither can you.

When it comes to your microbiome, it should be safety first.

Government agencies in country after country around the world are using global institutions such as the United Nations and World Health Organization to try to stem the tide of the NCD epidemic. At the same time such agencies realize that their past and present safety testing programs have not covered the microbiome. They have been misdirected and are currently of questionable use. Watch to see the different government responses. Those countries making decisions that err on the side of your safety are more likely to be supporting foods, environments, and lifestyles that support the health of all of you, the superorganism.

Consider government responses in the case of bisphenol A (BPA), the endocrine disruptor in baby bottles, food and drink packaging, medical tubing, and other plastics, which damages the immune and reproductive systems and, in doing so, contributes to the risk of NCDs. A 2010 publication in the *Proceedings of the National Academy of Sciences* demonstrated that bisphenol A compromises the gut barrier, which is the key interface connecting our microbiome to our immune system. Not surprisingly, this causes misregulated inflammation. Which countries banned it quickly, and which are having to be dragged into actively protecting health at a vulnerable life stage, infancy? In the case of BPA both the timing and extent of bans were revealing. Canada's ban took effect in 2010, Europe's in 2011, South America's (specifically Brazil, Argentina, and Ecuador) in 2012, and the United States' partial ban began in 2013 after much consumer protest. That tells you who is in the forefront of environmental health concerns and who lags.

Here is another case to consider that is still in progress. As of March 2015, the NIH reported their research findings showing common food emulsifiers disrupt the gut microbiome and provide a pathway to NCDs, including inflammation-driven obesity. The biology seems clear on this toxic effect. How long will it take before precautions are taken to protect consumers? Which regulatory groups in which countries will act, and which won't? While some may argue that we don't want knee-jerk reactions to initial research results, the past few decades show us the health implications of foot-dragging when chemical toxicities and biological plausibility are clear. That is why we are seeing the ever-increasing rise in NCDs related to immune dysfunction. We are allowing ourselves to become incomplete humans.

The same issues relate to current uncertainties and the opportunities for consumers to make their own choices and control what goes into or on their own bodies. Which countries are proactive in helping consumers see the composition of their food, including food additives and modifiers, and which countries seem unconcerned? As a toxicologist, I say you have a right to know. Read the list of ingredients. Pay attention to the chemical components of your environment. Agencies and governments will follow. In a way, we are all toxicologists now. Canaries are notoriously sensitive to change in their environment. If the canary in the coal mine died, it signaled that the miners' health was in jeopardy from odorless toxic gases. The microbiome is our new personal canary. Changes in the microbiome measured by changes in the metabolites in our breath or urine or the abundance of certain bacteria in feces can be an early warning of unhealthy chemical exposures.

We already know that some components of processed food (emulsifiers) and common drugs such as NSAIDs can be toxic for the microbiome at doses previously thought to be safe. But most chemicals and drugs have not been examined for toxicity to our microbiome. For those that have, the microbiome-drug interactions are extensive.

Government agencies have handled the uncertainty of toxicity in various ways. In the United States the burden of proof usually falls on whether something has been shown to be toxic beyond the shadow of a doubt—innocent until proven guilty, so to speak. In contrast, in Europe a different standard exists. This is known as the precautionary principle. Under this principle the default is that, if human harm is plausible, exposing large numbers of people to a new chemical or drug is considered too great a risk until it has been proven that the exposure is safe. So a suspect chemical or drug is held back until safety is established. The most celebrated decision under the precautionary principle concerns GMO foods. The precautionary principle has led Europe to block or move very slowly on GMO crops and foods, with each crop evaluated individually, whereas the reverse has been true in the US.

Regardless of your personal opinion on GMO safety, the research into our microbiome and the regulations surrounding it are a very different kind of issue. We know far more about genes and how they function than we know about the behavior of our microbial selves. We all have personal choices to make, but if anything, more caution seems warranted when it comes to the microbiome.

The herbicide glyphosate is at the center of the current debate about insecticides and herbicides, where regulation has focused on the safety of human mammalian cells. Glyphosate is not only the active component of a herbicide product produced by Monsanto, but is also part of the GMO strategy of producing glyphosate-resistant crops to grow in soil with ever-increasing concentrations of glyphosate. What does glyphosate do in soil? Among other things it selectively alters the ecosystem of environmental microbes, such as favoring the formation of some types of biofilms. A recent report from South America indicates that glyphosate reduces the presence of fungi symbiotic to grass roots that are needed to support foraging animals. Also, glyphosate can change things like antibiotic susceptibility or resistance in types of bacteria that are pathogenic to humans. The world is increasingly becoming a glyphosate-rich planet.

Amazingly, there are only a few studies examining how this massively distributed chemical affects the microbiome, and most of these studies are on food-producing animals. Nevertheless, they reveal reasons for concern. In a study of chicken gut microbes, researchers found that pathogenic bacteria were more resistant to glyphosate than were the helpful commensal bacteria.

In Germany, glyphosate shifted the mix of microbes in cows; in particular it reduced the natural protection provided by commensal bacteria in restricting the growth of botulism spores. There has been an increase in botulism-related disease in cattle in recent years, and glyphosate causes the problem by suppressing the antagonistic effect of *Enterococcus* species on *Clostridium botulinum*. We have a massive amount of a new chemical dumped into both the environment and the food chain without fully understanding the potential risk to our microbiome.

If your government doesn't exercise due caution when it comes to the prevention of NCDs, then you may need to become proactive in avoiding foods, chemicals, and drugs when you aren't satisfied with the level of ingredient and safety information that is provided.

Healthy Choices

Obesity, diabetes, heart disease, cancer, Alzheimer's disease, dementia, autism, Parkinson's disease, celiac disease, inflammatory bowel disease, and hundreds more NCDs are the new normal. They kill us, change our lives, make many of us invalid and ever more dependent upon others, drain our health services and pocketbooks, and in some cases stress government resources to the limits. We haven't found the solution to date. But then again, we had hardly begun to understand the problem until now.

Having introduced the completed self hypothesis as a fundamental principle of a healthier life, in this final chapter I will briefly pull together the initiatives that can help you to seed, feed, and protect a robust, diversified microbiome. The fact that there is more than one diet and microbial mix that supports a healthy life adds complexity to your choices. But it also means you are not searching for a single fountain of youth. You are simply looking for one of several options, for a lifestyle that supports your personal ecosystem with its thousands of cohabitating species. It matters less what works for your neighbor than what works for your body.

Diet is connected to the microbiome, which is in turn connected back to dietary cravings. Both are connected to inflammation, which can be locked into place by a dysfunctional microbiome. Exercise can help lessen the risk of NCDs. Nevertheless, early-life programming and epigenetic factors can block even well-conceived attempts at later-life solutions. Single-factor solutions and single-life-stage views of the risk of NCDs are likely to be less successful than looking at all the ways you can support your whole body across the entire life span. Diet alone, exercise alone, and probiotics alone are all less likely to work than an integrated and comprehensive effort to support the whole you.

For many people, diet alone simply doesn't work. If you are among those where diets did not take, it is not your fault or a lack of willpower. Brute-force diets don't work for many of us because our microbes either fail to shift enough, or they don't shift fast enough. The microbes will call for the foods (energy sources) that sustain them, and they know exactly how to make you crave those foods. In effect, a majority part of you is working against your well-intentioned diet. Having the wrong mix of microbes is like trying to hammer a square peg into a round hole. There is a slight chance you can do it, but you might well damage everything in the process. Piecing together a healthy diet and installing the microbes in you that want that healthy diet is a much better strategy.

Note that in many countries probiotics are considered to be medical foods and should be taken under the guidance of health care professionals. It is important to work on diet and those things that metabolize the diet to the benefit of your whole superorganism. Probiotics such as those in many fermented foods have been consumed as a personal choice for centuries, yet in many ways the same microbes are regulated much like a drug if consumed outside the fermented food itself. You choose your meals three times a day without professional advice, but taking probiotics falls into a gray area. Consider the most commonly used probiotic found in many yogurts, *Lactobacillus acidophilus*. Is it a food if it exists naturally in food? Is it a food if a supplement is added to fortify it? Is it automatically a drug if taken separately from the food, or is it still a food supplement?

These issues affect both what manufacturers can claim about probiotics and how the FDA handles them. As recently noted by the University of Maryland Medical Center, the US FDA has yet to approve *Lactobacillus acidophilus* for any medical use, even though some health care practitioners may recommend it. In fact, making disease claims about probiotics not approved as a new drug will get a company shut down.

Each of us needs to take some initiative, but the road to a healthy microbiome gradually requires less effort. Apart from simply feeling better, when your microbiome is better balanced, it will not only accept but crave good healthy food. Meanwhile, here are ten wide-ranging key suggestions for becoming a healthier superorganism based on the new biology. They are healthy choices.

1. First and foremost, do not delay seeding a baby's microbiome. Have a plan for it, even if cesarean delivery is necessary or elected. Vaginal swabs may offer one option. While installing a complete microbiome can be helpful at any life stage, including at my age, it is most effective early in life.

2. Breast milk has been designed through centuries of crafting to feed the whole baby, including the baby's microbiome. While it may be plausible to design equivalent substitutes that provide everything in breast milk, including the maternal immune factors, those perfect substitutes do not exist at present. If a substitute for breast milk is needed (e.g., enhanced formula), feeding the baby's microbiome should be part of the plan.

3. In the event you need antibiotics, discuss complementary probiotic therapy with your health professionals. In general, it is good to check if there is safety information relevant to the microbiome for any prescribed or over-the-counter drug. A hope is that the database on drug-microbiome interactions will increase significantly in the near future.

4. Prebiotics are important. For adjusting your microbiome to reduce the risk of NCDs or reduce the inflammation that supports these diseases, supplementation with prebiotics is as important as taking probiotics. You need to provide the food for the microbes you want to have. It is their action in using the prebiotics that alters your metabolism and provides useful signals that reach your immune system, gut, brain, liver, endocrine organs, and other tissues.

5. Pay attention to how your body reacts to drugs as well as chemicals in your environment, including household and personal care products. Unless someone can show you data that the drugs and chemicals are safe for your microbiome, they don't actually know if they are truly safe for you. Trust your body and its responses to exposure.

6. If members of your family are not predisposed to multiple allergies via overproduction of IgE antibodies (a condition known as atopy) and you plan to have a furry pet, it is better to have the pet in the household as early in a baby's life as possible. There is reduced risk of animal-associated asthma during childhood. Plus, a dog in the household will actually help exchange microbes among family members.

7. If you have food intolerances or allergies, you need to work carefully around those in any rebiosis effort to install a new microbiome and feed it. Some people have seen their intolerances go away, but that is more likely after you have been able to shift your microbiome.

8. Fecal microbiota transplantation is a major alteration where both the procedure itself and the selection of the donor are very important. It needs to be performed under the supervision of health professionals and only in cases of absolute necessity.

9. If you have food cravings and would like to change them, rebiosis using probiotics can work. Food preferences, including cravings, change as your gut microbes shift and begin to exert their influence on your brain.

10. Mood swings, anxiety, brain fog, and depression have often been attributed to hormone imbalances. But we now know that your microbiome is the master controller of neurobehavioral changes. Working on that via rebiosis has the potential to spare you decades of heavy-duty medications, each with its own cadre of side effects.

These ten initiatives are a good start. But let us dig a little more into the details of some of the things you can take action on now.

Probiotics

If the microbiome is severely altered from a healthy balance, consumption of a single probiotic strain is unlikely to bring everything back into balance. For this reason, many clinical trials have used combinations of probiotics along with prebiotics and even antioxidants to attack inflammation. NCDs such as obesity are pro-inflammatory conditions where the healthier ratio of higher Bacteroidetes to lower Firmicutes has been reversed, with Firmicutes becoming the more predominant group. The goal of rebiosis, including consumption of probiotics, is to reverse the imbalance.

Probiotics aren't just things you swallow. They have been used in the nose, mouth, and vagina, and topically on skin. Tablets containing *Lactobacillus reuteri* have been effective in reducing inflammation associated with periodontal disease. *Lactobacilli* and, in particular, those related to *L. acidophilus* (*L. crispatus, L. gasseri,* and *L. jensenii*), are useful for making the vagina more acidic and protecting against bacterial vaginosis. *L. plantarum* has been used for respiratory tract priming. *L. rhamnosus* has been used in nasal priming to prevent damaging inflammation.

Rebiosis and protection of the skin microbiome is in its infancy. But a wealth of probiotic and prebiotic products are likely to become available soon to enhance skin health. Both oral and topical applications of probiotics have been studied. In mice, oral intake of *Bifidobacterium breve* B-3 protected the skin of the animals from photo damage by ultraviolet radiation. *Lactobacillus plantarum* HY7714 had a similar effect, preventing the skin from drying out and thickening after UV exposure.

In a human pilot study, a new probiotic cosmetic containing the skin bacterium *Staphylococcus epidermidis* was tested and found to enhance the lipid content of skin and keep the skin from drying out. Such things as microbially based sunscreens and cosmetics are on the horizon.

Fermented Foods

According to a 2006 World Health Organization definition, probiotics are live microbes that can promote health when given in appropriate amounts. They can comprise a single microbial species or a mixture. Usually, these are ingested with the idea of altering gut microbe makeup. This may occur as the ingested microbes take up residence in specific areas of your gut. Alternatively, some microbes may simply spend time in the gut digesting their preferred foods. Through their brief presence and with chemicals they produce, they can also affect other microbes in the area and your own mammalian cells. The International Scientific Association for Probiotics and Prebiotics, whose membership spans academia, industry, and government, is a useful source of information about probiotics in foods and supplements.

In most cases, probiotic consumption has focused on ingesting microbe-containing foods or dietary supplements with the aim of altering the makeup and/or the function of our gut microbes. However, in theory, there could be probiotics for the skin, airways, or reproductive tract as well. You may soon see advertisements for reproductive probiotics. In this case, the probiotic is likely to be applied rather than ingested. Ironically, today probiotics are encountered by a majority of people as supplements to our diet. However, this was not always the case. For earlier generations, probiotic consumption was thought of as an integral part of the diet. Each geographic area and culture had its own source of food microbes. We know these foods today under the general name of fermented foods.

Ideas about the possible health benefits of what are now known as probiotics date back centuries. Even early human civilization took advantage of the health benefits of probiotics via the consumption of various fermented foods. Fermentation is a process in which an organism such as a bacterium or yeast takes up a carbohydrate as a desired food. This would often be a starch or a type of sugar. Via digestion of this food, the organism produces various waste products. This is usually an alcohol, an acid, or a gas. These specific waste products have the effect of making the food more acidic (lowering the pH). The change in pH and the environment of the food itself has a protective effect that prevents food from spoiling and stops the growth of pathogen-producing microbes. In the absence of refrigeration, fermented food could be stored longer and consumed safely. This was a massive benefit for ancient civilizations where food availability was often a life-or-death issue. Indigenous peoples of virtually every region on earth developed fermented foods to support life and health within their societies.

Fermentation was used by the Sumerians, Babylonians, ancient Chinese, Egyptians, Greeks, and Romans. The foods these ancient cultures fermented included bread, sauces (e.g., soy), dairy (milk and cheese), vegetables (such as cabbage, turnip, squash, and carrots), alcoholic beverages (e.g., wine and beer), meats (e.g., sausage), and chocolate. Fermented foods we see today include amasi, ayran, various cheeses, sauerkraut, kimchi, miso, pickled herring, poi, soy sauce, sourdough bread, torshi, yogurt, tempeh, and traditionally made cod-liver oil. Most regions of Africa have multiple types of fermented foods that have been handed down from their ancestors. Each food is tailored to the produce that could be grown in that specific region.

If fermented foods are not a part of your regular diet, you are not alone. Consumption of these foods has waned as westernized processed foods have grown in impact. An additional fact is that the national food guides published by government and medical agencies are virtually devoid of fermented foods. Researchers have speculated that, because many of these foods have been prepared in homes rather than via large-scale production through major commercial food companies, fermented foods have simply been off the radar when it comes to recommended nutrition.

In fact, a group of research scientists in Canada recently examined both the traditions of fermented foods across virtually every culture and the exclusion of them from virtually every government-based nutritional guide. You won't find them in the food pyramids of the United States (via the USDA), China, or Japan. One exception seems to be India, where fermented foods have been recommended for pregnant women. Occasionally, yogurts may be mentioned in some guidelines. This is something that needs to change as we shift from a focus solely on the mammalian human to a more complete view of the feeding and care of our superorganism. We should seize every safe opportunity to support our superorganism.

There is a distinction between fermented foods, pickled foods, and foods with live probiotic cultures. Fermented foods have the benefits of containing probiotic bacteria and/or yeast. The microbes may or may not be alive at the time they are consumed. The microbes grew in the food and metabolized food components to produce microbial metabolites. For example, the activity of the bacteria in dairy products can reduce lactose, thus making the product more digestible for some individuals. However, if the "fermented" foods have been pasteurized, this kills the microbial cultures. No live probiotic bacteria will be ingested if the food has been excessively heated. Therefore, it is useful to examine exactly what the label says in terms of food processing to determine if the food contains live, active cultures. Of course, when possible, preparation of fermented foods at home is one way to know how a food was handled. It will most likely be more economical as well. A short list of some popular fermented foods is described below.

1. Sauerkraut

I have some personal experience with fermented foods, although until recently it was the lack thereof. The one most closely connected to my family is sauerkraut. I should note that while sauerkraut is most closely associated with German culture, pickled cabbage dishes existed in ancient Rome and were a staple of the armies of Genghis Khan and the Tartars. Its adoption in Germany was probably later. Dutch and English ship captains took sauerkraut with them on voyages as a long-lasting food staple and for protection against scurvy. In addition to gathering fresh fruit whenever he could, British naval captain James Cook took thousands of pounds of sauerkraut with him on his almost three-year-long voyage launched in 1768 to circle to globe. Reports suggest that the regular sailors resisted eating it until Cook ordered it served daily at the captain's table and for officers. Variations on pickled cabbage recipes are known in France as choucroute and in Russia as shchi. Some recent efforts are under way to standardize sauerkraut production with specific bacterial isolates of *Lactobacillus plantarum* and *Leuconostoc mesenteroides*.

My own family's experience with sauerkraut mirrors what happened across different communities and cultures. My father's side of the family was part of a migratory wave from Germany to the Texas hill country in the early 1850s. Sauerkraut was a staple. Fast-forward to the twentieth century, when my great-grandfather and great-grandmother were both fluent in German and, when excited, my dad's oma would slip into her ancestral language. Sauerkraut, along with a special cookie, was core to their diet.

In the next generation, my grandfather still spoke German and, in his youth, was a US Army motorcade driver during World War I. He was occasionally asked to help translate when German troops were encountered along the roads. Later, he was a San Antonio city councilman. Again sauerkraut was prevalent in this household, which included my father. While my father occasionally requested it as a food from his youth, he hadn't learned how to cook it, and my mother (Scots-Irish background) had little experience preparing it. Additionally, she would always comment on the days of lengthy preparation and lingering kitchen odors after any attempt. As a result I grew up with a generally negative view of this traditional stinky food. If you ever encountered it, you know store-bought sauerkraut of the 1960s was not pouring off the shelves for good reason. It was nothing like homemade. This parallels the experience of my wife, who, as a little girl, ate sauerkraut once a week when her father had it on Saturdays with hot dogs. After his death, this food and that tradition were gone from the family. One of our friends had a similar experience growing up in California in the mid-twentieth century as sauerkraut waned as the older generation passed. Through similar experiences, traditional fermented foods can become lost from our diets.

2. Kimchi

Kimchi is a Korean staple that was originally made from cabbage but has since included other vegetables as base ingredients. It is so popular, so much a part of Korean history, and so connected to health promotion that the Korea Tourism Organization site promotes it as one of the reasons to visit Korea. In addition to a high fiber content, it contains several vitamins and minerals and, of course, the live *Lactobacillus* bacteria that help in the fermentation process. Among these is one named for the dish itself, *Lactobacillus kimchii*, but it includes *Leuconostoc* and *Weissella* bacteria as well. Among the numerous variations of kimchi, consumption tends to vary by region of the country and season of the year. There is even a museum devoted to kimchi located in Seoul.

3. Kombucha

In the beverage category, kombucha is a sweetened black tea with a slight fizz that has been around for thousands of years. Green tea versions have been developed as well. The beverage contains multiple species of both bacteria and yeast and goes under different names in different countries. It is thought to have originated somewhere in northern Asia (possibly China). Most preparations contain a pancake-like film on top called a SCOBY (symbiotic culture of bacteria and yeast), with the effervescing, slightly sour liquid underneath.

Kombucha is known to contain several B vitamins and, depending on the length of fermentation, some vitamin C. While it has been used for a variety of ailments for centuries, actual health impact remains to be firmly established. Additionally, some caution is appropriate to avoid excessive stomach acid and risk of ulcers or allergic reactions. The beverage has become available even in discount chains, and in some eateries is sold on tap. Refermentation in certain bottled preparations also has been a concern since it could lead to higher-than-expected alcohol content. Various positive health effects have been attributed to its consumption.

4. Miso

Miso is among the oldest probiotic-containing fermented foods. It originated in Japan as a salt-containing, fermented, soybean-based food. The microbial part of miso is a fungus called *Aspergillus oryzae*. The metabolism of some chemicals in soy by the fungus is thought to convey some of the reported benefits of miso. Antioxidants are among these fungal metabolites. Some versions of miso add barley and other grains as well as other microorganisms in the fermentation process. The food is often sold as a refrigerated paste to be used as a condiment in soups and other foods. To preserve the active cultures in miso and avoid killing off the active microorganisms with heat, miso is often added to foods at the end of cooking or even as foods are cooling. Among the properties recently attributed to miso is the degradation of histamine.

5. Tempeh

Tempeh originated in Indonesia and is made from whole beans that are fermented. It was originally made from soybeans, though some modern-day recipes substitute other types of beans. Tempeh is started with a fungus, *Rhizopus oligosporus,* that produces a natural antibiotic that works against certain gut pathogens. The food is also high in antioxidant properties.

6. Kvass

Kvass is a Russian beverage that was traditionally made using a type of sourdough bread, resulting in something akin to beer but without the alcohol content. However, there are many variations on this that replace the grains with fruits or vegetables. Among the most popular is beet kvass. As a probiotic source, this is excellent in that beets provide useful nutrients plus their own sugar, and the fermentation process adds the probiotic bacteria. All that is needed then is salt and water. Often whey is added as well to affect the type of bacteria that grow in the mix. Among the bacteria associated with kvass are *Lactobacillus casei, Leuconostoc mesenteroides,* and *Saccharomyces cerevisiae.*

7. Amasi

Africa has its own varieties of fermented foods. Among these is the probiotic milk drink called amasi. It emerged among traditional people of several African countries (e.g., Zimbabwe, Kenya, South Africa). While the name may be in common across country borders, the exact mix of bacteria in the cultures can differ. However, lactic-acid-producing bacteria appear to lead the way among the microbial mixtures. One of the suggested benefits is that consumption of amasi reduces the prevalence of diarrheal diseases. This may occur through protection against certain pathogenic strains of *E. coli* bacteria. In analysis of amasi, *Lactococcus lactis* was the predominant isolate, with several species from *Leuconostoc* and *Enterococcus* also prevalent.

8. Chicha

Central and South America also have a fermented foods tradition. In Peru, the Inca drank a type of corn beer made with maize called chicha de jora. Other areas of Central and South America used different base foods such as cassava, potatoes, quinoa, rice, and pineapple. A recent analysis of chicha found numerous bacterial genera represented (*Lactobacillus, Bacillus, Leuconostoc, Enterococcus, Streptomyces, Enterobacter, Acinetobacter, Escherichia, Cronobacter, Klebsiella*).

How would a microbiome-supportive diet differ from generally healthy diets? It might not differ at all. A microbiome-supportive diet is likely to include some fermented foods as a natural way to supply both probiotics and microbial metabolites. It would include prebiotics designed to promote the growth and revival of your helpful microbial co-partners. These would include fiber and sugars that are indigestible by your mammalian parts but feed your specific microbes. Prebiotics would include things like (1) resistant starches such as those as found in raw bananas, (2) inulin, a group of polysaccharides, found in onion, chicory, garlic, and asparagus, (3) fructo-oligosaccharides, many of which are breakdown products of inulin, and (4) galacto-oligosaccharides, which are breakdown products of lactose. In a human trial with stressed college students, the administration of a prebiotic galacto-oligosaccharide supplement was found to reduce gastrointestinal diseases and the duration and severity of colds.

Exercise

You have probably heard that exercise is good for you. It supports your heart and circulation as well as immune and neurological functions. There are numerous programs designed to encourage a lifestyle with regular exercise beginning with children in schools, continuing as adults enter the workplace, and persisting among the elderly as well. With other aspects of the microbiome, we are only beginning to appreciate how environmental conditions, including lifestyle, can affect the composition and diversity of our co-partners.

The level of exercise seems to affect the ratio of Bacteroidetes to Firmicutes in the gut, as well as microbial production of metabolites such as short-chain fatty acids (butyrate, acetate, propionate). Too little exercise can be associated with an unhealthy microbial balance and/or a leaky gut, allowing bacterial metabolites into areas of the body in concentrations that enhance rather than dampen inflammation. Overexercise with serious calorie restriction (creating something similar to anorexia) restricts the number of useful gut microbes. A happy medium of food intake and exercise level lies between the extremes, helping us maintain an immune-balanced, anti-inflammatory mix of personal microbes.

Some researchers have described effects of exercise on the microbiome that are independent of diet. In fact, when you are working through the microbiome, exercise can blunt the adverse effects of a high-fat diet. In research using rodents, voluntary versus forced exercise (which is a type of stressor) produced opposite impacts on the microbiome as well as on the levels of inflammation, with voluntary exercise being more anti-inflammatory. With forced exercise, rodents are made to run on a treadmill rather than allowed to choose to use it. That stresses them out, negating the otherwise useful benefits of the physical movement, and hurts the microbiome. The difference might be analogous to your choosing to use the gym's treadmill or spend an evening swing dancing versus having to run across London or New York City in a rainstorm because you missed your bus and can't get a taxi. The latter involves physical movement but is unlikely to help your microbiome. Finally, in a mouse study, exercise was shown to spare, in part, the toxic effects of polychlorinated biphenyl (PCB) exposure through its effects on the gut microbiome.

These findings indicate the importance of exercise as a way to maintain a healthy microbiome.

Go Be Completely Beautiful

Across this book's chapters I have discussed the completed self-hypothesis, the new biology, the revolution in medicine, and the opportunities for self-care that lead the way toward reversing the epidemic of NCDs. So much is changing as we embrace ourselves not simply as individuals but rather as a community of cooperative species, an entire ecosystem. But it is presenting us with a golden opportunity for improved health and well-being.

We are just now beginning to discover and appreciate the living organisms within us that make health and life possible. Yet it wasn't always so, and to illustrate just how much our views have changed, here's a vignette from the life of Alexander Fleming, discoverer of penicillin.

In the early 1930s, Charles Wilson, who was dean of the medical school at St. Mary's, where Fleming worked, had the opportunity to build a new school. As part of it, Almroth Wright, Fleming's mentor and boss, got a new Institute of Pathology attached to Wilson's school. In 1933, the new buildings were finished. They were so prestigious that, on

December 12, King George V and Queen Mary were on hand to open them. Wright wanted impressive displays for the inoculation department, and Fleming decided to get artistic. He created germ paintings using bacteria with different pigments.

To make his paintings, Fleming took sketches executed on absorbent cards or blotting paper and filled them in using chromogenic bacteria he'd grown on agar plates. Through a painstaking process, differently colored bacteria were used to create new works of art . . . He produced landscapes, ballerinas, guardsmen, and a fleur-de-lis in St. Mary's blue. Quite special above all of his works, and appropriate for this royal visit, he painted the Union Jack, the flag of the United Kingdom, using bacteria.

When Queen Mary came by and saw it, she was not amused and sniffed, "Yes—but what good is it?"

The beauty of the microbial world is a new discovery, yet it is ancient, and it is very much our own.

RESOURCES FOR PROBIOTICS

When it comes to probiotics and their food, prebiotics, there are several useful resources. The International Scientific Association for Probiotics and Prebiotics (ISAPP), an association of academic and industrial scientists, has a sole focus on probiotics and prebiotics and is the first place to look for the latest developments. They maintain a useful, up-to-date resource guide for both probiotics and prebiotics on their website: http://www.isapp.net/Probiotics-and-Prebiotics/Resources.

Additionally, their site links to a useful probiotics guide developed by Primary Care Network (a primary medical care educational organization): http://www.isapp.net/Portals/0/docs/News/merenstein%20sanders%20CME%20Probiotics.pdf. This guide describes the important considerations of strain and manufacturing source for probiotic products as well as the evidence for specific probiotics as applied to specific diseases. A recent guide was also developed for probiotics available in Canada (http://www.isapp.net/Portals/0/docs/clincial%20guide%20canada.pdf). These links and content information provide detailed background that you can use for working with your health professionals as you make self-care choices involving everything that goes into and on your body: food, prebiotics and probiotics, drugs, and consumer products.

Below is a list of current and anticipated probiotics described in recent studies and reviews:

Akkermansia muciniphila
Bifidobacterium animalis
Bifidobacterium breve
Bifidobacterium infantis
Bifidobacterium longum
Enterococcus durans
Enterococcus faecalis
Faecalibacterium prausnitzii
Lactobacillus acidophilus
Lactobacillus amylovorus
Lactobacillus casei
Lactobacillus fermentum
Lactobacillus gasseri
Lactobacillus helveticus
Lactobacillus johnsonii
Lactobacillus kefiranofaciens
Lactobacillus paracasei
Lactobacillus plantarum, Note: one of few used for gut and also
    used in respiratory tract priming
Lactobacillus reuteri
Lactobacillus rhamnosus GG
Pediococcus pentosaceus
Saccharomyces boulardii
VSL#3 blend

You have probably heard the term "heart-healthy diet," and the recommendations from government and private groups alike to consume a diet rich in
fruits, vegetables, and whole grains with a lower intake of meat and unhealthy
fats than is presently consumed in many Western countries. The diet of the
Mediterranean region serves as one model for a balance of these dietary components, as recommended by the Mayo Clinic. Other groups have published
dietary suggestions aimed at supporting physiological systems, balance, and
function. Examples of this would be the science-based recommendations of
(1) Ronald Watson, a nutritional immunologist and professor at the University
of Arizona and lead author of *Nutrition in the Prevention and Treatment of Abdominal Obesity*, (2) Tom Malterre, a nutritionist and coauthor of *The Elimina-*

tion Diet, and (3) Susan Prescott, a clinical immunologist and professor in the School of Paediatrics and Child Health, University of Western Australia, author of *Origins: Early-Life Solutions to the Modern Health Crisis.* Other groups such as the Neurological Health Foundation have published dietary recommendations for general physiology or focused on the brain and neurological systems. Unquestionably, poor diets can contribute to microbial dysbiosis and NCDs, while well-balanced healthy diets can support a well-balanced microbiome and health.

ACKNOWLEDGMENTS

This book would not have happened but for the vision, support, collaborative assistance, and expertise of many other individuals. If the human superorganism represents a cast of thousands, then this book on the superorganism has a cast of a least tens to hundreds. I extend thanks to all, even if many are unnamed. I remain forever grateful for those who helped steer me into and through this writing endeavor.

First and foremost, my thanks go to my wife, Janice. She kept telling me that I needed to write this book even when it seemed as if other projects were destined to block the way. In some ways, my primary role was merely to have the good sense to listen to my wife. Once the book was undertaken, Janice's knowledge base, editorial skills, and writing prowess were invaluable in translating my scientific jargon into something more readable. This book would not have been written but for Janice.

The editors at Dutton and, specifically, Stephen Morrow and Adam O'Brien, deserve special thanks. They believed in this project early on, provided the context and opportunity, and were ever ready with expert guidance that was instrumental in the development of this book. I am in awe of their professional expertise and marvel at the skill with which they pursue their craft.

I want to thank my former graduate students and postdoctoral researchers as well as numerous research collaborators who span many decades of my career. They contributed so much to our scientific knowledge base and were critical partners in my own scientific journey.

I have a deep appreciation for the culture of my department at Cornell, led by Dr. Avery August, and the entire scholarly environment of Cornell University. To paraphrase Ezra Cornell's vision, it is a place where anyone can come to learn anything. I feel like I have spent my academic career making the most of the educational richness that Cornell has to offer to all who set foot on our campus. For any future college students reading this book, if you are curious about life, Cornell is the place for you.

Finally, I want to thank my parents, extended family, and many teachers. They created an environment in which the entire world was open to me and I could pursue my dreams. I will never forget their role and the importance of simply allowing things to unfold.

NOTES

Introduction: The New Medical Landscape

2 **Koch's postulates** See early English translation: Koch R. *Investigations into the Etiology of Traumatic Infective Diseases.* London, England: The New Sydenham Society; 1880. http://pds.lib.harvard.edu/pds/view/7027406. Accessed June 24, 2015; Blevins SM, Bronze MS. Robert Koch and the "golden age" of bacteriology. *Int J Infect Dis.* 2010;14(9):e744–751. doi:10.1016/j.ijid.2009.12.003.

2 **during World War II** Oatman E. The drug that changed the world. *Journal of the College of Physicians & Surgeons of Columbia University.* Winter 2005;25(1). http://www.cumc.columbia.edu/psjournal/archive/winter-2005/drug.html. Accessed September 2, 2015.

3 **the White Death, and leprosy** Dormandy T. *The White Death—A History of Tuberculosis.* London, England: Hambledon Press; 1999.

3 **during the nineteenth and twentieth centuries** Koehler CW. Consumption, the great killer. *thetimeline. mdd.* 2002;5(2):47–49. http://pubs.acs.org/subscribe/archive/mdd/v05/i02/html/02timeline.html. Accessed August 30, 2015.

3 **waiting to die** Sucre R. The great white plague: the culture of death and the tuberculosis sanatorium. http://www.faculty.virginia.edu/blueridgesanatorium/death.htm. Accessed August 30, 2015.

3 **"Sanatorium, Texas"** Henderson JC. Sanatorium, TX. *Handbook of Texas Online.* 2010. https://tshaonline.org/handbook/online/articles/hls16. Accessed August 30, 2015.

3 **the 1860s and 1960s** Wong A. When the last patient dies. *The Atlantic.* May 27, 2015. http://www.theatlantic.com/health/archive/2015/05/when-the-last-patient-dies/394163/. Accessed August 30, 2015.

3 **families to stay together** Waters MF, Rees RJ, Pearson JM, et al. Rifampicin for lepromatous leprosy: nine years' experience. *Br Med J.* 1978;1(6106):133–136; Zaffiri L, Gardner J, Toledo-Pereyra LH. History of antibiotics: from salvarsan to cephalosporins. *J Invest Surg.* 2012;25(2):67–77.

3 decades of the twentieth century WGBH Educational Foundation and Vulcan Productions, Inc. Deadly diseases: polio. Rx for Survival. http://www.pbs
.org/wgbh/rxforsurvival/series/diseases/polio.html. Accessed August 31, 2015.

4 in the early 1950s Meier P. The biggest public health experiment ever: the 1954 field trial of the Salk poliomyelitis vaccine. In: Tanur JM, Mosteller F, Kruskal WH, et al. *Statistics: A Guide to the Unknown.* Belmont, CA: Duxbury; 1989. http://www.medicine.mcgill.ca/epidemiology/hanley/c622/salk_trial.pdf. Accessed August 31, 2015.

4 Basil O'Connor to run it Franklin D. Roosevelt. Whatever happened to polio? *Smithsonian National Museum of American History.* http://amhistory.si
.edu/polio/howpolio/fdr.htm. Accessed August 31, 2015.

4 March of Dimes March of Dimes. Whatever happened to polio? *Smithsonian National Museum of American History.* http://amhistory.si.edu/howpolio/fdr
.htm. Accessed August 31, 2015.

4 rarely reared its head since Meier P. The biggest public health experiment ever: the 1954 field trial of the Salk poliomyelitis vaccine. In: Tanur JM, Mosteller F, Kruskal WH, et al. *Statistics: A Guide to the Unknown.* Belmont, CA: Duxbury; 1989. http://www.medicine.mcgill.ca/epidemiology/hanley/c622/
salk_trial.pdf. Accessed August 31, 2015.

5 those cells are microbial Stein R. Finally, a map of all the microbes on your body. NPR. June 13, 2012. http://www.npr.org/sections/health-shots
/2012/06/13/154913334/finally-a-map-of-all-the-microbes-on-your-body. Accessed August 31, 2015; Sender R, Fuchs S, Milo R. Revised estimates for the numbers of human and bacteria cells in the body. *bioRxiv.* 2016. Preprint. January 6, 2016. doi:10.1101/036103. http://www.biorxiv.org/early/2016/01/06/036103
.full.pdf. Accessed January 14, 2016.

5 no single person carries them all Mundasad S. Human microbiome project reveals largest microbial map. *BBC News.* June 13, 2012. http://www.bbc
.com/news/health-18422288. Accessed August 30, 2015.

5 850 on the skin Bouslimani A, Porto C, Rath CM, et al. Molecular cartography of the human skin surface in 3D. *Proc Natl Acad Sci USA.* 2015;112(17), e2120–e2129; Hilty M, Burke C, Pedro H, et al. Disordered microbial communities in asthmatic airways. *PLoS ONE.* 2010;5(1):e8578. doi:10.1371/journal.pone
.0008578.

5 in the urogenital tract Hummelen R, Fernandes AD, Macklaim JM, et al. Deep sequencing of the vaginal microbiota of women with HIV. *PLoS ONE.* 2010;5(8):e12078. doi:10.1371/journal.pone.0012078; The Scientist Staff. The body's ecosystem. *The Scientist.* August 1, 2014. http://www.the-scientist
.com/?articles.view/articleNo/40600/title/The-Body-s-Ecosystem/. Accessed August 31, 2015.

5 3,000 square inches Weyrich LS, Dixit S, Farrer AG, et al. The skin microbiome: associations between altered microbial communities and disease. *Australas J Dermatol.* February 25, 2015. doi:10.1111/ajd.12253; Skin. *National Geographic.* http://science.nationalgeographic.com/science/health-and
-human-body/human-body/skin-article/. Accessed August 31, 2015.

5 your sweaty toes Norton A. Scientists map the fungi on your feet. *U.S. News Health.* May 22, 2013. http://health.usnews.com/health-news/news/articles
/2013/05/22/scientists-map-the-fungi-on-your-feet. Accessed August 31, 2015.

6 Carl Linnaeus Paterlini M. There shall be order: the legacy of Linnaeus in the age of molecular biology. *EMBO Rep.* 2007;8(9):814–816.

6 Stephen Jay Gould Gould SJ, *Ontogeny and Phylogeny.* Cambridge, MA: Harvard University Press; 1977.

7 23 percent of deaths The top 10 causes of death. *World Health Organization Fact Sheet.* http://who.int/mediacentre/factsheets/fs310/en/index2.html. Accessed August 31, 2015.

8 87 percent of all deaths The top 10 causes of death. *World Health Organization Fact Sheet.* http://who.int/mediacentre/factsheets/fs310/en/index2 .html. Accessed August 31, 2015.

9 68 percent of all deaths The top 10 causes of death. *World Health Organization Fact Sheet.* http://who.int/mediacentre/factsheets/fs310/en/index2 .html. Accessed August 31, 2015; Bloom DE, Caliero ET, Jane-Llopis E, et al. The global economic burden of noncommunicable diseases. Geneva: World Economic Forum; 2011. http://www.weforum.org/EconomicsOfNCD.

9 World Health Organization Global Status Report on NCDs. Chronic diseases and health promotion. *World Health Organization.* http://www.who .int/chp/ncd_global_status_report/en/. Accessed June 23, 2015.

9 United Nations Global Status Report on NCDs. Chronic diseases and health promotion. *World Health Organization.* http://www.who.int/nmh/events/un_ ncd_summit2011/political_declaration_en.pdf. Accessed June 23, 2015.

11 "incredibly selfish" Allen V. Girl, 4, went into anaphylactic shock and lost consciousness on a plane after selfish passenger ignored three warnings not to eat nuts on board. *Mail Online.* August 14, 2014. http://www.dailymail .co.uk/news/article-2724684/Nut-allergy-girl-went-anaphylactic-shock-plane -passenger-ignored-three-warnings-not-eat-nuts-board.html. Accessed June 18, 2015.

11 their country of origin Davidson NK, Moreland P. Living with diabetes blog: traveling with diabetes—plan ahead. *Mayo Clinic.* http://www.mayo clinic.org/diseases-conditions/diabetes/expert-blog/traveling-with-diabetes/ bgp-20056556. Accessed June 23, 2015.

11 stabilize blood sugar levels Davidson NK, Moreland P. Living with diabetes blog: traveling with diabetes—plan ahead. *Mayo Clinic.* http://www.mayo clinic.org/diseases-conditions/diabetes/expert-blog/traveling-with-diabetes /bgp-20056556. Accessed June 18, 2015.

12 it was 1 in 45 *Centers for Disease Control and Prevention.* http://www.cdc.gov/ nchs/data/nhsr/nhsr087.pdf. Accessed November 18, 2015.

13 to rise to 42 percent by 2030 Finkelstein EA, Khavjou OA, Thompson H, et al. Obesity and severe obesity forecasts through 2030. *Am J Prev Med.* 2012;42(6):563–570.

13 useless to him Cooperstein P. Obese man says airline made him book 2 tickets, then gave him seats 2 rows apart. *Business Insider.* October 14, 2013. http:// www.businessinsider.com/obese-man-says-airline-made-him-book-2-tickets -then-gave-him-seats-2-rows-apart-2013-10. Accessed June 23, 2015.

14 "on any of [their] aircraft" Grenoble R. Airline flies obese man to destination— then refuses to fly him home. *Huff Post Travel.* November 9, 2013. http://www .huffingtonpost.com/2013/11/09/too-fat-to-fly-british-airways-kevin-chenais _n_4242407.html. Accessed June 23, 2015.

14 lack of exercise Engber D. Fat-E: the new Pixar movie goes out of its way to equate obesity with environmental collapse. *Slate.* July 10, 2008. http://www.slate.com/articles/health_and_science/green_room/2008/07/fate.html. Accessed June 23, 2015.

14 as child abusers Luscombe R. Puerto Rico law would brand parents of obese children "child abusers." *The Guardian.* March 8, 2015. http://www.theguardian.com/world/2015/mar/08/puerto-rico-childhood-obesity-law-parents-abuse. Accessed June 23, 2015.

15 safely entertain guests Cable A. Whiff of perfume could kill me. *Mail Online.* http://www.dailymail.co.uk/health/article-299652/Whiff-perfume-kill-me.html. Accessed June 28, 2015.

16 encouraging others to follow Indoor environmental quality policy. Office of Health and Safety. *Centers for Disease Control and Prevention.* http://www.drsteinemann.com/Resources/CDC%20Indoor%20Environmental%20Quality%20Policy.pdf. Accessed June 23, 2015.

16 fragrance-free workplace policy Scent-free policy for the workplace, OSH Answers Fact Sheets. *Canadian Centre for Occupational Health and Safety.* http://www.ccohs.ca/oshanswers/hsprograms/scent_free.html. Updated June 24, 2015. Accessed June 23, 2015.

16 many popular perfumes Wendlandt A. What's in a scent? Perfume makers adapt to EU rules. *Reuters.* July 7, 2104. http://www.reuters.com/article/2014/07/07/us-perfume-regulation-insight-idUSKBN0FC0EB20140707. Accessed June 23, 2015.

16 Marilyn Monroe Tadeo M. Iconic Chanel No 5 perfume to reformulate under new EU regulations. *Independent.* May 29, 2014. http://www.independent.co.uk/news/business/news/iconic-chanel-no-5-perfume-to-reformulate-under-new-eu-regulations-9451331.html. Accessed June 18, 2015.

16 didn't change; we did Tadeo M. Iconic Chanel No 5 perfume to reformulate under new EU regulations. *Independent.* May 29, 2014. http://www.independent.co.uk/news/business/news/iconic-chanel-no-5-perfume-to-reformulate-under-new-eu-regulations-9451331.html. Accessed June 18, 2015.

1. The End of the Old Biology

22 genetics and human nature Dobzhansky TG. *The Biological Basis of Human Freedom.* New York, NY: Columbia University Press; 1956.

23 "gene machines" Dawkins R. *The Selfish Gene.* Oxford, England: Oxford University Press; 1976.

24 North America, Europe, and Asia The human genome project completion: frequently asked questions. National Human Genome Research Institute. *NIH.* https://www.genome.gov/11006943. Accessed June 23, 2015.

25 journal *Nature* Lander ES, Linton LM, Birren B, et al. Initial sequencing and analysis of the human genome. *Nature.* 2001;409(6822):860–921.

25 approximately 22,000 genes Cutter AD, Dey A, Murray RL. Evolution of the *Caenorhabditis elegans* genome. *Mol Biol Evol.* 2009;26(6):1199–1234.

28 2012 paper I wrote Dietert RR, Dietert J. The completed self: an immunological view of the human-microbiome superorganism and risk of chronic diseases. *Entropy.* 2012;14(11):2036–2065. doi:10.3390/e14112036.

28 **"fills me with astonishment"** Dawkins R. *The Selfish Gene*. Oxford, England: Oxford University Press; 1976.

2. Superorganism Ecology

32 **"public health and basic science"** Relman DA. Microbiology: learning about who we are. *Nature*. 2012;486(7402):194–195.

33 **cause malaria** Malaria: malaria parasites. *Centers for Disease Control and Prevention*. http://www.cdc.gov/malaria/about/biology/parasites.html. Accessed June 6, 2015.

35 **equally represented** Ter Steege H, Pitman NCA, Sabatier D, et al. Hyperdominance in the Amazonian tree flora. *Science*. 2013;342(6156). doi:10.1126/science.1243092.

36 **species of butterflies** Rainforests: exotic, diverse and highly threatened. *The Nature Conservancy*. http://www.nature.org/ourinitiatives/habitats/rainforests/index.htm. Accessed June 7, 2015.

36 **multiple scientific societies** Ochwang'i DO, Kimwele CN, Oduma JA, et al. Medicinal plants used in treatment and management of cancer in Kakamega County, Kenya. *J Ethnopharmacol*. 2014;151(3):1040–1055.

36 **natural antimicrobials** Qin S, Li J, Chen H-H, et al. Isolation, diversity, and antimicrobial activity of rare actinobacteria from medicinal plants of tropical rain forests in Xishuangbanna, China. *Appl Environ Microbiol*. 2009;75(19):6176–6186.

36 **rosy periwinkle** Chivian E, Bernstein A, eds. *Sustaining Life: How Human Health Depends on Biodiversity*. Oxford, England: Oxford University Press; 2008.

36 **tropical rain forests** Lang SS. Tom Eisner, "father of chemical ecology" and renowned Cornell biologist, dies at 81. *Cornell Chronicle*. March 27, 2011. http://www.news.cornell.edu/stories/2011/03/tom-eisner-father-chemical-ecology-dies-81. Accessed June 7, 2015.

36 **Eisner's sensibility** Ghazoul J, Sheil D. *Tropical Rain Forest Ecology, Diversity, and Conservation*. Oxford, England: Oxford University Press; 2010.

38 **diversity as well** Newton AC, ed. *Biodiversity Loss and Conservation in Fragmented Forest Landscapes*. Wallingford, England: CABI; 2007.

38 **research team** Mueller RC, Paula FS, Mirza BS, et al. Links between plant and fungal communities across a deforestation chronosequence in the Amazon rainforest. *ISME J*. 2014;8(7):1548–1550.

38 **photosynthesis by the algae** Barott KL, Venn AA, Perez SO, et al. Coral host cells acidify symbiotic algal microenvironment to promote photosynthesis. *Proc Natl Acad Sci USA*. 112(2):607–612.

39 **urchins, and clams** What species live in and around coral reefs? *National Ocean Service*. http://oceanservice.noaa.gov/facts/coral_species.html. Accessed June 7, 2015.

39 **biologist David Mindell** Mindell DP. Phylogenetic consequences of symbioses: Eukarya and Eubacteria are not monophyletic taxa. *Biosystems*. 1992;27(1):53–62. doi:10.1016/0303-2647(92)90046-2.

39 **degradation of the coral reef** Pollock FJ, Lamb JB, Field SN, et al. Sediment and turbidity associated with offshore dredging increase coral disease prevalence on nearby reefs. *PLoS ONE*. 2014;9(7):e102498. doi:10.1371/journal.pone.0102498.

40 **weedy, large algae** Littler MM, Littler DS, Brooks BL. Harmful algae on tropical coral reefs: bottom-up eutrophication and top-down herbivory. *Harmful Algae.* 2006;5(5):565–585.

40 **bacteria, fungi, and viruses** Blaser MJ. The microbiome revolution. *J Clin Invest.* 2014;124(10):4162–4165.

41 **1,000 different species of bacteria** Grice EA, Kong HH, Conlan S, et al. Topographical and temporal diversity of the human skin microbiome. *Science.* 2009;324(5931):1190–1192.

41 **harbor approximately 1,000 different bacterial species** Qin J, Li R, Raes J, et al. A human gut microbial gene catalogue established by metagenomic sequencing. *Nature.* 2010;464(7285):59–65.

41 **minority of those genes** Li J, Jia H, Cai X, et al. An integrated catalog of reference genes in the human gut microbiome. *Nat Biotechnol.* 2014;32(8):834–841.

41 **quarter of their genes** Greenblum S, Carr R, Borenstein E. Extensive strain-level copy-number variation across human gut microbiome species. *Cell.* 2015;160(4):583–594.

41 **consortium of researchers** Bouslimani A, Porto C, Rath CM, et al. Molecular cartography of the human skin surface in 3D. *Proc Natl Acad Sci USA.* 2015;112(17):e2120–e2129.

41 **small intestine or skin** The Scientist Staff. The body's ecosystem. *The Scientist.* August 1, 2014. http://www.the-scientist.com/?articles.view/article No/40600/title/The-Body-s-Ecosystem/. Accessed July 13, 2015.

42 **"biome depletion"** Beales DL. Biome depletion in conjunction with evolutionary mismatches could play a role in the etiology of neurofibromatosis 1. *Med Hypotheses.* 2015;84(4):305–314.

43 **importance to the ecosystem** Mouillot D, Bellwood DR, Baraloto C, et al. Rare species support vulnerable functions in high-diversity ecosystems. *PLoS Biol.* 2013;11(5):e1001569. doi:10.1371/journal.pbio.1001569.

44 **epithelial and immune cells** Everard A, Belzer C, Geurts L, et al. Cross-talk between *Akkermansia muciniphila* and intestinal epithelium controls diet-induced obesity. *Proc Natl Acad Sci USA.* 2013;110(22):9066–9071; Owens B. Gut microbe may fight obesity and diabetes. *Nature News.* 2013. doi:10.1038/nature.2013.12975.

44 **human mammalian species** Gordon HA, Pesti L. The gnotobiotic animal as a tool in the study of host microbial relationships. *Bacteriol Rev.* 1971;35(4):390–429.

44 **germ-free mice** Williams SCP. Gnotobiotics. *PNAS USA.* 2014;111(5):1661.

45 **in order to survive** Gordon HA, Pesti L. The gnotobiotic animal as a tool in the study of host microbial relationships. *Bacteriol Rev.* 1971;35(4):390–429.

45 **three days and die** Hirayama K, Uetsuka K, Kuwabara Y, Tamura M, Itoh K. Vitamin K deficiency of germfree mice caused by feeding standard purified diet sterilized by gamma-irradiation. *Exp Anim.* 2007;56(4):273–278.

45 **immune responses** Gordon HA, Pesti L. The gnotobiotic animal as a tool in the study of host microbial relationships. *Bacteriol Rev.* 1971;35(4):390–429; Heneghan J, ed. *Germfree Research: Biological Effect of Gnotobiotic Environments.* New York, NY: Academic Press; 1973; Umesaki Y. Use of gnotobiotic mice to identify and characterize key microbes responsible for the development of the intestinal immune system. *Proc Jpn Acad Ser B Phys Biol Sci.* 2014;90(9):313–332.

45 **made by the microbiome** A brief history of the use of microfloras in gnotobiotic rodents. *Taconic*. http://www.taconic.com/prepare-your-model/pre
conditioning-solutions/microbiome-solutions/brief-history-of-the-use-of
-microfloras-in-gnotobiotic-rodents.html. Accessed June 6, 2015.

45 **gut bacteria are killed off** Shirakawa H, Komai M, Kimura S. Antibiotic-
induced vitamin K deficiency and the role of the presence of intestinal flora.
Int J Vitam Nutr Res. 1990;60(3):245–251.

3. The Invisible Human Superorganism

46 **ten million microbial genes** Li J, Jia H, Cai X, et al. An integrated catalog of
reference genes in the human gut microbiome. *Nat Biotechnol*. 2014;32(8):834–
841; INRA News Office. Cataloguing 10 million human gut microbial genes: an
unparalleled accomplishment. *INRA Science & Impact*. 2014. http://www
.inra.fr/en/Scientists-Students/Mechanisms-of-living-organisms/All-the-news
/Cataloguing-10-million-human-gut-microbial-genes. Accessed July 14, 2015.

48 **very long time** Weiner MS. *The Rule of the Clan: What an Ancient Form of Social Organization Reveals About the Future of Individual Freedom*. New York, NY:
Farrar, Straus and Giroux; 2013.

48 **continues even today** Schatz E. *Modern Clan Politics: The Power of "Blood" in
Kazakhstan and Beyond*. Seattle, WA: University of Washington Press; 2004.

48 ***Missing Microbes*** Blaser MJ. *Missing Microbes: How the Overuse of Antibiotics
Is Fueling our Modern Plagues*. New York, NY: Henry Holt & Co; 2014.

49 **what was required** Dietert RR, Dietert J. Impact of women on the history of
Scottish goldsmithing. *History Scotland Magazine*. 2011;11(6):48–53.

49 **wealth and property** Narayan A. Matrilineal society. *Encyclopaedia Britan-
nica*. http://www.britannica.com/topic/matrilineal-society. Accessed July 3,
2015.

49 **80 percent are patriarchal** Divale WT, Harris M. Population, warfare, and
the male supremacist complex. *Am Anthropol*. 1976;78(3):521–538.

49 **patrilineal lineage** Holden CJ, Sear R, Mace R. Matriliny as daughter-biased
investment. *Evol Hum Behav*. 2003;24(2):99–112.

50 **chapel and a padded cell** Claire. The birth of Elizabeth I. *The Anne Boleyn
Files*. 2010. http://www.theanneboleynfiles.com/the-birth-of-elizabeth-i/. Accessed July 2, 2015.

52 **fifty-five million by the year 2020** Tsai T. China has too many bachelors.
PRB. 2012. http://www.prb.org/Publications/Articles/2012/china-census
-excess-males.aspx. Accessed July 2, 2015; Poston DL, Conde E, DeSalvo B. China's unbalanced sex ratio at birth, millions of excess bachelors and societal
implications. *Vulnerable Child Youth Stud*. 2011;6(4):314–320.

52 **increased violence against women** Trivedi A, Timmons H. India's man
problem. India Ink. *The New York Times*. January 16, 2013. http://india.blogs
.nytimes.com/2013/01/16/indias-man-problem/. Accessed July 2, 2015.

52 **enforcement has been problematic** Hesketh T, Lu L, Xing ZW. The consequences of son preference and sex-selective abortion in China and other Asian
countries. *CMAJ*. 2011;183(12):1374–1377.

53 **that of the mouth** Aagaard K, Ma J, Antony KM, et al. The placenta harbors a
unique microbiome. *Sci Transl Med*. 2014;6(237):237ra65.

53 ***Lactobacillus*** **bacteria** Zheng J, Xiao X, Zhang Q, et al. The placental micro-

biome varies in association with low birth weight in full-term neonates. *Nutrients.* 2015;7(8):6924–6937. doi:10.3390/nu7085315.

54 ***Bacteroides,* and *Clostridium*** Adlerberth I, Wold AE. Establishment of the gut microbiota in Western infants. *Acta Paediatr.* 2009;98(2):229–238.

54 **carried against the mother's skin** Dietert RR. Natural childbirth and breastfeeding as preventive measures of immune-microbiome dysbiosis and misregulated inflammation. *J Anc Dis Prev Rem.* 2013;1:103. doi:10.4172/jadpr .1000103.

54 **with few exceptions** Dietert RR. Natural childbirth and breastfeeding as preventive measures of immune-microbiome dysbiosis and misregulated inflammation. *J Anc Dis Prev Rem.* 2013;1:103. doi:10.4172/jadpr.1000103.

54 **could harm the baby** Grandjean P, Satoh H, Murata K, Eto K. Adverse effects of methylmercury: environmental health research implicatons. *Environ Health Perspect.* 2010;118(8):1137–1145.

54 **needed by our microbes** Marcobal A, Barboza M, Froehlich JW, et al. Consumption of human milk oligosaccharides by gut-related microbes. *J Agric Food Chem.* 2010;58(9):5334–5340.

54 **stages of an infant's life** Chichlowski M, German JB, Lebrilla CB, Mills DA. The influence of milk oligosaccharides on microbiota of infants: opportunities for formulas. *Annu Rev Food Sci Technol.* 2011;2:331–351.

55 **baby will consume** McGuire MK, McGuire MA. Human milk: Mother Nature's prototypical probiotic food? *Adv Nutr.* 2015;6(1):112–123.

55 **bacteria in breast milk** Soto A, Martin V, Jiménez E, et al. Lactobacilli and bifidobacteria in human breast milk: influence of antibiotherapy and other host and clinical factors. *J Pediatr Gastroenterol Nutr.* 2014;59(1):78–88.

55 **elective C-section** Cabrera-Rubio R, Mira-Pascual L, Mira A, Collado MC. Impact of mode of delivery on the milk microbiota composition of healthy women. *J Dev Orig Health Dis.* 2016;7(1):54–60. doi:10.1017/S2040174415001397.

55 **residence in the baby's gut** Rautava S. Early microbial contact, the breast milk microbiome and child health. *J Dev Orig Health Dis.* 2016;7(1)5–14; McGuire MK, McGuire MA. Human milk: Mother Nature's prototypical probiotic food? *Adv Nutr.* 2015;6(1):112–123.

56 **prior publications** Dietert RR. Developmental immunotoxicology (DIT): windows of vulnerability, immune dysfunction and safety assessment. *J Immunotoxicol.* 2008;5(4):401–412.

56 **affected by the microbiome** Dietert RR, Etzel RA, Chen D, et al. Workshop to identify critical windows of exposure for children's health: immune and respiratory systems work group summary. *Environ Health Perspect.* 2000;108(Suppl 3):483–490; Olszak T, An D, Zeissig S, et al. Microbial exposure during early life has persistent effects on natural killer T cell function. *Science.* 2012;336(6080):489–493; Principles of pediatric environmental health: why are children often especially susceptible to the adverse effects of environmental toxicants? *Environ Health Med Edu. ATSDR.* 2012. http://www.atsdr.cdc.gov/ csem/csem.asp?csem=27&po=3. Accessed September 4, 2015.

56 **helping the other out** Lodwig EM, Hosie AHF, Bourdès A, et al. Amino-acid cycling drives nitrogen fixation in the legume-*Rhizobium* symbiosis. *Nature.* 2003;422(6933):722–726.

57 **upper atmosphere** DeLeon-Rodriguez N, Lathem TL, Rodriguez-RLM, et al. Microbiome of the upper troposphere: species composition and prevalence,

effects of tropical storms, and atmospheric implications. *Proc Natl Acad Sci USA.* 2013;110(7):2575–2580.

57 **Latin into "clean"** Vaishampayan P, Moissl-Eichinger C, Pukall R, et al. Description of *Tersicoccus phoenicis* gen. nov., sp. nov. isolated from spacecraft assembly clean room environments. *Int J Syst Evol Microbiol.* 2013;63(Pt 7):2463–2471.

57 **ultraviolet radiation** Vaishampayan PA, Rabbow E, Horneck G, Venkateswaran KJ. Survival of *Bacillus pumilus* spores for a prolonged period of time in real space conditions. *Astrobiology.* 2012;12(5):487–497.

57–58 **humanoid communities** Tito RY, Knights D, Metcalf J, et al. Insights from characterizing extinct human gut microbiomes. *PLoS ONE.* 2012;7(12):e51146. doi: 10.1371/journal.pone.0051146.

58 **bacteria found in the mouth** Adler CJ, Dobney K, Weyrich LS, et al. Sequencing ancient calcified dental plaque shows changes in oral microbiota with dietary shifts of the Neolithic and Industrial revolutions. *Nat Gen.* 2013;45(4):450–455.

59 **nucleus have mitochondria** Sagan L. On the origin of mitosing cells. *J Theor Biol.* 1967;14(3):255–274.

59 *Acquiring Genomes* Margulis L, Sagan D. *Acquiring Genomes: A Theory of the Origins of Species.* New York, NY: Basic Books; 2002.

59 **archaea and bacteria** Tekle YI, Parfrey LW, Katz LA. Molecular data are transforming hypotheses on the origin and diversification of Eukaryotes. *Bioscience.* 2009;59(6):471–481.

59 **our bacterial ancestors** Crisp A, Boschetti C, Perry M. Expression of multiple horizontally acquired genes is a hallmark of both vertebrate and invertebrate genomes. *Genome Biol.* 2015;16(1):50. doi:10.1186/s.13059015-0607-3.

60 **William Cullen** Cullen WR. *Is Arsenic an Aphrodisiac?: The Sociochemistry of an Element.* London, England: Royal Society of Chemistry Publishing; 2008.

61 **and others** Gustafson RF. The Upas tree: Pushkin and Erasmus Darwin. *PMLA.* 1960;75(1):101–109.

62 **function of the microbiome** Dietert RR, Silbergeld EK. Biomarkers for the 21st century: listening to the microbiome. *Toxicol Sci.* 2015;144(2):208–216.

62 **called xenobiotics** Maurice CF, Haiser HJ, Turnbaugh PJ. Xenobiotics shape the physiology and gene expression of the active human gut microbiome. *Cell.* 2013;152(1–2):39–50; Carmody RN, Turnbaugh PJ. Host-microbial interactions in the metabolism of therapeutic and diet-derived xenobiotics. *J Clin Invest.* 2014;124(10):4173–4181.

63 **2010 Vancouver Winter Olympics** Kurczy S. Lindsey Vonn race suit: how big an advantage at Vancouver Olympics? *The Christian Science Monitor.* February 26, 2010. http://www.csmonitor.com/World/Olympics/Olympics-blog/2010/0226/Lindsey-Vonn-race-suit-How-big-an-advantage-at-Vancouver-Olympics. Accessed October 10, 2015.

63 **wear them in competition** *Daily Mail* Reporter. Unsuitable: Team USA drop their high-tech speedskating suits after disappointing opening week at the Olymipcs. *Mail Online.* February 14, 2014. http://www.dailymail.co.uk/news/article-2559755/The-secret-suits-High-tech-speedskating-uniforms-created-space-engineers-main-complaint-shocking-U-S-performance-Sochi.html. Accessed October 10, 2015.

63 **suits and/or equipment** Chadwick N. Will success on the ice of Sochi rekin-

dle enthusiasm for an Amsterdam Olympics? *The Amsterdam Herald.* March 1, 2014. http://www.amsterdamherald.com/index.php/election-2012-blog/1159 -20140301-will-success-on-ice-of-sochi-rekindle-enthusiasm-amsterdam- olympics. Accessed October 10, 2015.

64 three sets of triplets Murphy K, O'Shea CA, Ryan CA, et al. The gut microbiota composition in dichorionic triplet sets suggests a role for host genetic factors. *PLoS ONE.* 2015;10(4):e0122561. doi:10.1371/journal.pone.0122561.

65 our puppet, Pinocchio Stilling RM, Dinan TG, Cryan JF. The brain's Geppetto-microbes as puppeteers of neural function and behaviour? *J Neurovirol.* 2015. doi:10.1007/S13365-015-0355x.

4. The Incomplete Generation

66 congenital heart defects Kim JH, Scialli AR. Thalidomide: the tragedy of birth defects and the effective treatment of disease. *Toxicol Sci.* 2011;122(1):1–6.

66 sensory challenges Green CR, Roane J, Hewitt A, et al. Frequent behavioural challenges in children with fetal alcohol spectrum disorder: a needs-based assessment reported by caregivers and clinicians. *J Popul Ther Clin Pharmacol.* 2014;21(3):e405–e420.

67 "functional ability of the baby" Birth defects: specific birth defects. *Centers for Disease Control and Prevention.* http://www.cdc.gov/ncbddd/birth defects/types.html. Accessed July 3, 2015.

67 physiological changes Birth defects research. The Teratology Society. http:// www.teratology.org/index.asp. Accessed July 3, 2015.

68 degenerative disorders What are the types of birth defects? Eunice Kennedy Shriver National Institute of Child Health and Human Development. *NIH.* https://www.nichd.nih.gov/health/topics/birthdefects/conditioninfo/Pages/ types.aspx. Accessed July 4, 2015.

68 "how the body works" Birth defects & other health conditions. *March of Dimes.* http://www.marchofdimes.org/complications/birth-defects-and -health-conditions.aspx. Accessed July 4, 2015.

68 "environmentally induced" Birth defect. *Merriam-Webster.* http://www .merriam-webster.com/dictionary/birth%20defect. Accessed July 4, 2015.

68 several health problems Orme CM, Boyden LM, Choate KA, Antaya RJ, King BA. Capillary malformation—arteriovenous malformation syndrome: review of the literature, proposed diagnostic criteria, and recommendations for management. *Pediatr Dermatol.* 2013;30(4):409–415. doi:10.1111/pde.12112.

69 birth defect in the baby Dietert RR. The microbiome in early life: self-completion and microbiota protection as health priorities. *Birth Defects Res B Dev Reprod Toxicol.* 2014;101(4):333–340.

69 type of birth defect Dietert RR. The microbiome in early life: self-completion and microbiota protection as health priorities. *Birth Defects Res B Dev Reprod Toxicol.* 2014;101(4):333–340.

70 propionic acid MacFabe DF, Cain NE, Boon F, et al. Effects of the enteric bacterial metabolic product propionic acid on object-directed behavior, social behavior, cognition, and neuroinflammation in adolescent rats: relevance to autism spectrum disorder. *Behav Brain Res.* 2011;217(1):47–54; MacFabe DF. Short-chain fatty acid fermentation products of the gut microbiome: implications in autism spectrum disorders. *Microb Ecol Health Dis.* 2012;23:19260;

MacFabe DF. The propionic acid rodent model of autism. 2012. Video. *Autism Canada.* https://vimeo.com/79418104.

70 risk of autoimmune disease Alenghat T. Epigenomics and the microbiota. *Toxicol Pathol.* 2015;43(1):101–106; An D, Oh SF, Olszak T, et al. Sphingolipids from a symbiotic microbe regulate homeostasis of host intestinal natural killer T cells. *Cell.* 2014;156(1–2):123–133.

70 interdisciplinary research team Powell S. Antibiotics, infants and food allergy. *University of South Carolina.* 2014. http://www.sc.edu/uofsc/posts/2014/07_love_bryan_infant_antibiotics.php#.VoErCbYrKW9. Accessed June 22, 2015.

71 food allergy epidemic Love B. Food allergies antibiotic exposure. *South Carolina College of Pharmacy.* 2014. https://www.sccp.sc.edu/content/cpos -research-feature-bryan-love-%E2%80%93-food-allergies-antibiotic-exposure. Accessed September 4, 2015.

72 leading cause of disability The U.S. government and global non-communicable diseases. Global Health Policy. *The Henry J. Kaiser Family Foundation.* 2014. http://kff.org/global-health-policy/fact-sheet/the-u-s -government-and-global-non-communicable-diseases/. Accessed June 21, 2015.

72 their action plans Noncommunicable diseases. Pan American Health Organization. *World Health Organization.* http://www.paho.org/hq/index.php ?option=com_content&view=article&id=771&Itemid=852. Accessed June 21, 2015.

72 partner initiatives Non communicable diseases and disability: creating synergies, reducing inequalities, advancing development. 2013. *NCD Alliance.* http://ncdalliance.org/sites/default/files/rfiles/NCDs%20and%20Disability% 20Final_0.pdf. Accessed June 21, 2015.

73 an atypical human body DePoy E, Gilson S. Disability as microcosm: the boundaries of the human body. *Societies.* 2012;2(4):302–316. doi:10.3390/ soc2040302.

73 from all microbes "Bubble boy" 40 years later: look back at heartbreaking case. *CBS News.* http://www.cbsnews.com/pictures/bubble-boy-40-years -later-look-back-at-heartbreaking-case/. Accessed July 5, 2015.

74 virus in the donor cells Stewart DD, Morgenstern J. *The Boy in the Plastic Bubble.* Movie. 1976. http://www.imdb.com/title/tt0074236/. Accessed July 5, 2015.

75 requiring many services from others Bebinger M. Troubled future for young adults on autism spectrum. *WBUR's CommonHealth Reform and Reality.* 2014. http://commonhealth.wbur.org/2014/06/aging-out-autism-services. Accessed September 4, 2015.

75 living directly with parents Adult life planning—housing. *Asperger/Autism Network.* http://www.aane.org/about_asperger_syndrome/adult_life_planning _housing.html. Accessed September 4, 2015; Kendall C. Asperger's syndrome in adults—living with your adult child. *Asperger's Syndrome Newsletter.* http:// www.aspergerssociety.org/aspergers-syndrome-in-adults-living-with-your-adult -child-90/. Accessed September 4, 2015.

75 with Asperger's syndrome Monsebraaten L. The autism project: York University students with Asperger's thrive in mentorship program. *The Star.* November 17, 2012. http://www.thestar.com/news/investigations/2012/11/17/ the_autism_project_york_university_students_with_aspergers_thrive_in_ mentorship_program.html. Accessed September 4, 2015.

75 **disabled student population** Individuals with Disabilities Education Act cost impact on local school districts. *Atlas.* 2015. http://atlas.newamerica.org/individuals-disabilities-education-act-cost-impact-local-school-districts. Accessed September 4, 2015.

76 **long-term caretakers** Stone RI, Wiener JM. Who will care for us? Addressing the long-term care workforce crisis. *The Urban Institute.* 2001. http://aspe.hhs .gov/basic-report/who-will-care-us-addressing-long-term-care-workforce-crisis. Accessed July 4, 2015.

77 **student services offices** Settlement agreement between the United States of America and Lesley University, DJ 202-36-231. January 2013. http://www.ada .gov/lesley_university_sa.htm. Accessed June 21, 2015.

77 **accommodation plans** Trotch C. Food for thought: applying the ADA to students with food allergies. *Natl Assoc of College and Univ Attorneys.* 2014;12(7). http://www.higheredcompliance.org/resources/FoodAllergies.pdf. Accessed June 21, 2015; Grasgreen A. Dining disabilities. *Inside Higher Ed.* January 2, 2013. https://www.insidehighered.com/news/2013/01/02/lesley-settlement-flags-food -allergies-and-campus-dining. Accessed June 21, 2015.

77 **improper immune maturation** Stefka AT, Feehley T, Tripathi P, et al. Commensal bacteria protect against food allergen sensitization. *Proc Natl Acad Sci USA.* 2014;111(36):13145–13150.

77 **and wealth** Nightingale CH. *Segregation: A Global History of Divided Cities.* Chicago, IL: University of Chicago Press; 2012; Bishop B. *The Big Sort: Why the Clustering of Like-Minded America Is Tearing Us Apart.* New York, NY: Houghton Mifflin Harcourt; 2008.

78 **unintentionally increase segregation** Antony PJ. *Segregation Hurts: Voices of Youth with Disabilities and Their Families in India.* Rotterdam, the Netherlands: Sense Publishers; 2013; Gates B, ed. *Learning Disabilities: Towards Inclusion.* 5th ed. London, England: Churchill Livingstone; 2007.

78 **still a challenge** NPR Staff. Learning with disabilities: one effort to shake up the classroom. NPR on WSKG Public Radio. 2014. http://www.npr.org/2014/04/27/307467382/learning-with-disabilities-one-effort-to-shake-up-the -classroom. Accessed June 22, 2015.

5. Gene Swaps and Switches

82 ***Corynebacterium diphtheriae*** Freeman VJ. Studies on the virulence of bacteriophage-infected strains of *Corynebacterium diphtheriae. J Bacteriol.* 1951;61(6):675–688.

82 **the same viruses** Akiba T, Koyama K, Ishiki Y, et al. On the mechanism of the development of multiple-drug-resistant clones of Shigella. *Jpn J Microbiol.* 1960;4(2):219–227.

82 **occasionally exchange genes** Liu L, Chen X, Skogerbø G, et al. The human microbiome: a hot spot of microbial horizontal gene transfer. *Genomics.* 2012;100(5):265–270.

82 **transfer of genes could occur** De la Cruz F, Davies J. Horizontal gene transfer and the origin of species: lessons from bacteria. *Trends Microbiol.* 2000;8(3):128–133; Salzberg SL, White O, Peterson J, Eisen JA. Microbial genes in the human genome: lateral transfer or gene loss? *Science.* 2001;292(5523):1903–

1906; Kurland CG, Canback B, Berg OG. Horizontal gene transfer: a critical view. *Proc Natl Acad Sci USA.* 2003;100(17):9658–9662.

82 **demonstrated in 2005** Monroe, D. Jumping genes cross plant species boundaries. *PLoS Biol.* 2006;4(1):e35. doi:10.1371/journal.pbio.0040035.

83 **unique enzyme activities** Crisp A, Boschetti C, Perry M, et al. Expression of multiple horizontally acquired genes is a hallmark of both vertebrate and invertebrate genomes. *Genome Biol.* 2015;16:50.

84 **control mechanism is a central** Crisp A, Boschetti C, Perry M, et al. Expression of multiple horizontally acquired genes is a hallmark of both vertebrate and invertebrate genomes. *Genome Biol.* 2015;16:50.

85 **epigenetic gene switches** Ginder GD. Epigenetic regulation of fetal globin gene expression in adult erythroid cells. *Transl Res.* 2015;165(1):115–125.

85 **hemoglobin expression** Steliou K, Boosalis MS, Perrine SP, et al. Butyrate histone deacetylase inhibitors. *Biores Open Access.* 2012;1(4):192–198.

86 **beta-thalassemia** Bianchi N, Chiarabelli C, Zuccato C, et al. Erythroid differentiation ability of butyric acid analogues: identification of basal chemical structures of new inducers of foetal haemoglobin. *Eur J Pharmacol.* 2015;752:84–91.

86 **personal book of life** Hoffmann A, Zimmermann CA, Spengler D. Molecular epigenetic switches in neurodevelopment in health and disease. *Front Behav Neurosci.* 2015;9:120. doi:10.3389/fnbeh.2015.00120.

87 **maintenance of memories** Zovkic IB, Guzman-Karlsson MC, Sweatt JD. Epigenetic regulation of memory formation and maintenance. *Learn Mem.* 2013;20(2):61–74.

87 **your immune response** van den Elsen PJ, van Eggermond MC, Wierda RJ. Epigenetic control in immune function. *Adv Exp Med Biol.* 2011;711:36–49. doi:10.1007/978-1-4417-8216-2_4.

87 **responses to those hormones** Wojcicka A, Piekielko-Witkowska A, Kedzierska H, et al. Epigenetic regulation of thyroid hormone receptor beta in renal cancer. *PLoS ONE.* 2014;9(5):e97624. doi:10.1371/journal.pone.0097624; Ouni M, Belot MP, Castell AL, et al. The P2 promoter of the IGF1 gene is a major epigenetic locus for GH responsiveness. *Pharmacogenomics J.* 2015. doi:10.1038/tpj .2015.26.

87 **quality of sperm production** Guerrero-Bosagna C, Savenkova M, Hague MM, et al. Environmentally induced epigenetic transgenerational inheritance of altered Sertoli cell transcriptome and epigenome: molecular etiology of male infertility. *PLoS ONE.* 2013;8(3):e59922. doi:10.1371/journal.pone.0059922; Meikar O, Da Ros M, Kotaja N. Epigenetic regulation of male germ cell differentiation. *Subcell Biochem.* 2013;61:119–138.

89 **developing immune system occurs** Dietert RR, Dietert J. *Strategies for Protecting Your Child's Immune System.* Hackensack, NJ: World Scientific Publishing; 2010; Dietert RR. Developmental immunotoxicity, perinatal programming, and noncommunicable diseases: focus on human studies. *Adv Med.* 2014. Article ID 867805. doi:10.1155/2014/867805.

89 **absence of key microbes** Murgatroyd C, Spengler D. Epigenetics of early child development. *Front Psychiatry.* 2011;2(16):1–15. doi:10.3389/fpsyt .2011.00016; Godfrey KM, Costello PM, Lillycrop KA. The developmental environment, epigenetic biomarkers and long-term health. *J Dev Orig Health Dis.*

2015;6(5):399–406; Lillycrop KA, Burdge GC. Maternal diet as a modifier of offspring epigenetics. *J Dev Orig Health Dis.* 2015;6(2):88–95; Zhao Q, Hou J, Chen B, et al. Prenatal cocaine exposure impairs cognitive function of progeny via insulin growth factor II epigenetic regulation. *Neurobiol Dis.* 2015;82:54–65; Dietert RR. Transgenerational epigenetics of endocrine-disrupting chemicals. In: Tollefsbol T, ed. *Transgenerational Epigenetics.* San Diego, CA: Elsevier; 2014:239–254; Remely M, Aumueller E, Jahn D, et al. Microbiota and epigenetic regulation of inflammatory mediators in type 2 diabetes and obesity. *Benef Microbes.* 2014;5(1):33–43; Interlandi J. The toxins that affected your great-grandparents could be in your genes. *Smithsonian Magazine.* 2013. http://www.smithsonianmag.com/innovation/the-toxins-that-affected-your-great-grandparents-could-be-in-your-genes-180947644/. Accessed July 13, 2015.

89 **maturity as you age** Dietert RR, Etzel RA, Chen D, et al. Workshop to identify critical windows of exposure for children's health: immune and respiratory systems work group summary. *Environ Health Perspect.* 2000;108(Suppl 3):483–490; Jernigan TL, Baaré WF, Stiles J, Madsen KS. Postnatal brain development: structural imaging of dynamic neurodevelopmental processes. *Prog Brain Res.* 2011;189:77–92; Harding R, Pinkerton KE, eds. *The Lung: Development, Aging and the Environment.* 2nd ed. San Diego, CA: Elsevier; 2014.

89 **developmental basis of heart disease** Barker DJ, Osmond C, Golding J, et al. Growth in utero, blood pressure in childhood and adult life, and mortality from cardiovascular disease. *BMJ.* 1989;298(6673):564–567; Barker DJ. The fetal and infant origins of adult disease. *BMJ.* 1990;301(6761):1111; Barker DJ. Fetal origins of cardiovascular disease. *Ann Med.* 1999;31(Suppl 1):3–6.

90 **Barker hypothesis** Paneth N, Susser M. Early origin of coronary heart disease (the "Barker hypothesis"). *BMJ.* 1995;310(6977):411–412.

90 **later-life disease** Fox DA, Lucchini R, Aschner M, et al. Local effects and global impact in neurotoxicity and neurodegeneration: the Xi'an International Neurotoxicology Conference. *Neurotoxicology.* 2012;33(4):629–630; Walker CL, Ho SM. Developmental reprogramming of cancer susceptibility. *Nat Rev Cancer.* 2012;12(7):479–486; Heindel JJ, Vandenberg LN. Developmental origins of health and disease: a paradigm for understanding disease cause and prevention. *Curr Opin Pediatr.* 2015;27(2):248–253; Bousquet J, Anto JM, Berkouk K, et al. Developmental determinants in non-communicable chronic diseases and ageing. *Thorax.* 2015;70(6):595–597.

90 **and colleagues** Crews D, Gillette R, Miller-Crews I, et al. Nature, nurture and epigenetics. *Mol Cell Endocrinol.* 2014;398(1–2):42–52.

90 **"understanding of biological processes"** Crews D, Gillette R, Miller-Crews I, et al. Nature, nurture and epigenetics. *Mol Cell Endocrinol.* 2014;398(1–2):42–52.

91 **chemicals in mice** Skinner MK, Savenkova MI, Zhang B, et al. Gene bionetworks involved in the epigenetic transgenerational inheritance of altered mate preference: environmental epigenetics and evolutionary biology. *BMC Genomics.* 2014;15:377. doi:10.1186/1471-2164-15-377.

91 **hunger winter** Lumey LH, Van Poppel FWA. The Dutch Famine of 1944–45: mortality and morbidity in past and present generations. *Soc Hist Med.* 1994;7(2):229–246. doi:10.1093/shm/7.2.229.

91 **Great Chinese Famine of 1958–61** Song S. Identifying the intergenerational effects of the 1959–1961 Chinese Great Leap Forward Famine on infant mortality. *Econ Hum Biol.* 2013;11(4):474–487; Pembrey M, Saffery R, Bygren LO. Human transgenerational responses to early-life experience: potential impact on development, health and biomedical research. *J Med Genet.* 2014;51(9):563–572.

92 metabolism later in life Tobi EW, Goeman JJ, Monajemi R, et al. DNA methylation signatures link prenatal famine exposure to growth and metabolism. *Nat Commun.* 2014;5:5592. doi:10.1038/ncomms6592.

92 diabetes as adults Van Abeelen AF, Elias SG, Bossuyt PM, et al. Famine exposure in the young and the risk of type 2 diabetes in adulthood. *Diabetes.* 2012;61(9):2255–2260.

92 general population Veenendaal MV, Painter RC, De Rooij SR, et al. Transgenerational effects of prenatal exposure to the 1944–45 Dutch famine. *BJOG.* 2013;120(5):548–553.

92 Great Leap Forward Smil V. China's great famine: 40 years later. *BMJ.* 1999;319(7225):1619–1621; Li Wei, Yang DT. The great leap forward: anatomy of a central planning disaster. *J Polit Econ.* 2005;113(4):840–877.

92 metabolic syndrome Li Y, Jaddoe VW, Qi L, et al. Exposure to the Chinese famine in early life and the risk of metabolic syndrome in adulthood. *Diabetes Care.* 2011;34(4):1014–1018; Wang PX, Wang JJ, Lei YX, et al. Impact of fetal and infant exposure to the Chinese Great Famine on the risk of hypertension in adulthood. *PLoS ONE.* 2012;7(11):e49720. doi:10.1371/journal.pone.0049720.

92 schizophrenia Xu MQ, Sun WS, Liu BX, et al. Prenatal malnutrition and adult schizophrenia: further evidence from the 1959–1961 Chinese famine. *Schizophr Bull.* 2009;35(3):568–576.

92 anemia Shi Z, Zhang C, Zhou M, Zhen S, Taylor AW. Exposure to the Chinese famine in early life and the risk of anaemia in adulthood. *BMC Public Health.* 2013;13:904. doi:10.1186/1471-2458-13-904.

93 our mammalian genes Kumar H, Lund R, Laiho A, et al. Gut microbiota as an epigenetic regulator: pilot study based on whole-genome methylation analysis. *MBio.* 2014;5(6):e02113-e02114. doi:10.1128/mBio.02113-14; Khan S, Jena G. The role of butyrate, a histone deacetylase inhibitor in diabetes mellitus: experiental evidence for therapeutic intervention. *Epigenomics.* 2015;7(4):669–680.

93 control of the gene switches Alenghat T. Epigenomics and the microbiota. *Toxicol Pathol.* 2015;43(1):101–106.

6. Redirecting Precision Medicine

98 the twenty-first century Jain KK. Personalized medicine. *Curr Opin Mol Ther.* 2002;4(6):548–558.

98 go back several decades Pucheril D. The history and future of personalized medicine. *Managed Care.* 2011. http://www.managedcaremag.com/content/history-and-future-personalized-medicine. Accessed September 7, 2015.

98 personalize our medicine Novelli G. Personalized genomic medicine. *Intern Emerg Med.* 2010;5(Suppl 1):S81–S90.

98 National Cancer Institute Kaiser J. Obama gives East Room rollout to Precision Medicine Initiative. *Science*Insider. 2015. http://news.sciencemag.org/biology/2015/01/obama-gives-east-room-rollout-precision-medicine-initiative. Accessed September 7, 2015; Collins FS, Varmus H. A new initiative on precision medicine. *N Engl J Med.* 2015;372(9):793–795.

99 the individual patient National Research Council (US) Committee on a Framework for Developing a New Taxonomy of Disease. *Toward Precision Medicine: Building a Knowledge Network for Biomedical Research and a New Taxonomy of Disease.* Washington, DC: National Academies Press; 2011.

99 electronic health records Precision Medicine in Action. Public Health Genomics. *Centers for Disease Control and Prevention.* http://www.cdc.gov/genomics /public/features/precision_med.htm. 2015. Accessed September 7, 2015.

99 a big problem Rubin R. Precision medicine: the future or simply politics? *JAMA.* 2015;313(11):1089–1091.

100 next annual visit Dietert RR, Dietert JM. The microbiome and sustainable healthcare. *Healthcare.* 2015;3(1):100–129. doi:10.3390/healthcare3010100.

100 adjustments to your microbiome Brown E. Designer microbiome: MIT biologists program common gut bacteria. *Los Angeles Times.* July 11, 2015. http:// www.latimes.com/science/sciencenow/la-sci-sn-designer-microbiome-20150709 -story.html. Accessed July 15, 2015.

101 patient-doctor visit Goodman B. Use time with your doctor wisely. *Arthritis Foundation.* http://www.arthritis.org/living-with-arthritis/health-care/your -health-care-team/doctors-appointment-challenge.php. Accessed September 7, 2015.

101 unnecessary tests and referrals Goodman B. Use time with your doctor wisely. *Arthritis Foundation.* http://www.arthritis.org/living-with-arthritis/ health-care/your-health-care-team/doctors-appointment-challenge.php. Accessed September 7, 2015.

7. The Immune System Gone Wrong

104 path in life Dietert RR. Developmental immunotoxicity, perinatal programming, and noncommunicable diseases: focus on human studies. *Adv Med.* 2014. Article ID 867805. doi:10.1155/2014/867805.

105 one hundred, and the number is growing The cost burden of autoimmune disease: the latest front in the war on healthcare spending. AARDA. NCAPG. 2011. http://www.diabetesed.net/page/_files/autoimmune-diseases.pdf. Accessed July 18, 2015.

105 women more than men More than 850 medicines in development for diseases that disproportionately strike women. *PhRMA.* 2011. http://www.phrma.org/ media/releases/more-850-medicines-development-diseases-disproportionately -strike-women. Accessed July 19, 2015.

105 *DSM* grows with new entries American Psychiatric Associations Highlights of changes from DSM-IV-TR to DSM-5. *Amer Psych Pub.* http://www .dsm5.org/documents/changes%20from%20dsm-iv-tr%20to%20dsm-5.pdf. Accessed July 18, 2015; DSM-5: changes to the diagnostic and statistical manual of mental disorders. *Anxiety and Depression Association of America.* http://www.adaa.org/understanding-anxiety/DSM-5-changes. Accessed July 18, 2015.

106 the Peanut Brigade Isaacs H. *Jimmy Carter's Peanut Brigade.* Dallas, TX: Taylor Publishing Co.; 1977.

106 commemorative plates exist Emert J. Carter's peanut brigade recounts 1977 inauguration. *WALB News* 10. January 20, 2009. http://www.walb.com/

story/9704379/carters-peanut-brigade-recounts-1977-inauguration. Accessed September 7, 2015.

106 banned from some schools Chirbas K. Elk Grove school district to ban peanut products from elementary schools. *The Sacramento Bee.* August 1, 2013. http://www.sacbee.com/news/local/health-and-medicine/article2578315.html. Accessed September 7, 2015.

107 ban them from flights Ginsberg L. The end of peanuts on planes? Airlines face a battle over nut allergies. *Yahoo! Travel.* August 22, 2014. https://www.yahoo.com/travel/airlines-and-nut-allergies-95399512052.html. Accessed September 7, 2015.

108 features are quite different Dietert RR. Macrophages as targets of developmental immunotoxicity. *OA Immunology.* 2014;2(1):2; Bulger M, Palis J. Environmentally-defined enhancer populations regulate diversity of tissue-resident macrophages. *Trends Immunol.* 2015;36(2):61–62.

108 reaching your internal tissues Khazan O. The secret to a tattoo's permanence: the immune system. *The Atlantic.* July 22, 2014. http://www.theatlantic.com/health/archive/2014/07/the-real-reason-tattoos-are-permanent/374825/. Accessed July 19, 2015.

109 published in early 2010 Dietert RR, Dietert J. *Strategies for Protecting Your Child's Immune System.* Hackensack, NJ: World Scientific Publishing; 2010.

110 coauthored in 2009 Dietert RR, Dietert J. *Strategies for Protecting Your Child's Immune System.* Hackensack, NJ: World Scientific Publishing; 2010.

111 airways, urogenital region Dietert RR. Macrophages as targets of developmental immunotoxicity. *OA Immunology.* 2014;2(1):2.

112 we know it in mammals Cooper EL. Comparative immunology. *Integr Comp Biol.* 2003;43(2):278–280. doi:10.1093/icb/43.2.278; Jarosz J, Gliński Z. Earthworm immune responses. *Folia Biol (Krakow).* 1997;45(1–2):1–9.

112 functions when needed German N, Doyscher D, Rensing C. Bacterial killing in macrophages and amoeba: do they all use a brass dagger? *Future Microbiol.* 2013;8(10):1257–1264.

112 comparable innate immune systems Dvořák J, Mančíková V, Piži V, et al. Microbial environment affects innate immunity in two closely related earthworm species *Eisenia andrei* and *Eisenia fetida. PLoS ONE.* 2013;8(11):e79257. doi: 10.1371/journal.pone.0079257.

114 asbestos health litigations Castleman BI. *Asbestos: Medical and Legal Aspects.* 4th ed. Englewood Cliffs, NJ: Aspen Publishers; 1996.

115 crayons containing asbestos Bienkowski B. Is your child coloring with asbestos? *Scientific American.* July 8, 2015. http://www.scientificamerican.com/article/is-your-child-coloring-with-asbestos/. Accessed July 16, 2015.

115 risk was simply not known Cherrie JW, Tindall M, Cowie H. Exposure and risks from wearing asbestos mitts. *Part Fibre Toxicol.* 2005;2(5). doi: 10.1186/1743-897722-5.

115 solvent for cleaning glassware Benzene. TOXNET. http://toxnet.nlm.nih.gov/cgi-bin/sis/search/a?dbs+hsdb:@term+@DOCNO+35. Accessed July 18, 2015; Durkee J. *Cleaning with Solvents: Science and Technology.* Oxford, England: William Andrew; 2014.

115 risks become apparent Smith MT, Jones RM, Smith AH. Benzene exposure and risk of non-Hodgkin lymphoma. *Cancer Epidemiol Biomarkers Prev.*

2007;16(3):385–391. doi:10.1158/1055-9965.EPI-06-1057; Pyatt DW, Stillman WS, Irons RD. Hydroquinone, a reactive metabolite of benzene, inhibits NF-kappa B in primary human CD4+ T lymphocytes. *Toxicol Appl Pharmacol.* 1998;149(2):178–184.

115 **banned in the 1970s** Public health statement for carbon tetrachloride. *ATSDR Toxic Substances Portal.* August 2005. http://www.atsdr.cdc.gov/phs/phs.asp ?id=194&tid=35. Accessed July 17, 2015; US Environmental Protection Agency. IRIS toxicological review of carbon tetrachloride (external review draft). *EPA Environmental Assessment.* http://cfpub.epa.gov/ncea/cfm/recordisplay .cfm?deid=119546. Accessed July 17, 2015.

115 **separated into thin threads** Asbestos exposure and cancer risk. *National Cancer Institute.* http://www.cancer.gov/about-cancer/causes-prevention/ risk/substances/asbestos/asbestos-fact-sheet. Accessed July 16, 2015.

115–16 **production of vermiculite** Asbestos CAS ID #: 1332-21-4. *ATSDR Toxic Substances Portal.* http://www.atsdr.cdc.gov/substances/toxsubstance.asp? toxid=4. Accessed July 16, 2015.

116 **called alveolar macrophages** Nagai H, Toyokuni S. Biopersistent fiber-induced inflammation and carcinogenesis: lessons learned from asbestos toward safety of fibrous nanomaterials. *Arch Biochem Biophys.* 2010;502(1):1–7.

116 **antioxidative defenses** Dostert C, Petrilli V, Bruggen R, et al. Innate immune activation through Nalp3 inflammasome sensing of asbestos and silica. *Science.* 2008;320(5876):674–677.

116 **predictable outcomes** Chew SH, Toyokuni S. Malignant mesothelioma as an oxidative stress-induced cancer: an update. *Free Radic Biol Med.* 2015;86:166–178.

116 **two critical features** Nishimura Y, Maeda M, Kumagai-Takei N, et al. Altered functions of alveolar macrophages and NK cells involved in asbestos-related diseases. *Environ Health Prev Med.* 2013;18(3):198–204.

116 **cancers to survive better** Maeda M, Nishimura Y, Kumagai N, et al. Dysregulation of the immune system caused by silica and asbestos. *J Immunotoxicol.* 2010;7(4):268–278.

117 **cells providing the lining border** Magouliotis DE, Tasiopoulou VS, Molyvdas PA, et al. Airways microbiota: hidden trojan horses in asbestos exposed individuals? *Med Hypotheses.* 2014;83(5):537–540.

117 **medical journal** Marshall BJ, Warren JR. Unidentified curved bacilli in the stomach of patients with Pastritis and peptic ulceration. *Lancet.* 1984;1(8390):1311–1315; Pajares JM, Gisbert JP. *Helicobacter pylori*: its discovery and relevance for medicine. *Rev Esp Enferm Dig.* 2006;98(10):770–785.

118 **as a stomach resident** Blaser MJ. *Missing Microbes.* New York, NY: Henry Holt; 2014.

118 **Columbus in the New World** Castillo-Rojas G, Cerbón MA, López-Vidal Y. Presence of *Helicobacter pylori* in a Mexican pre-Columbian mummy. *BMC Microbiol.* 2008;8:119. doi:10.1186/1471-2180-8-119.

118 **hygiene hypothesis** Strachan DP, Taylor EM, Carpenter RG. Family structure, neonatal infection, and hay fever in adolescence. *Arch Dis Child* 1996;74:422–426. doi:10.1136/adc.74.5.422.

119 **called dendritic cells** Amedei A, Codolo G, Del Prete G, et al. The effect of *Heliobacter pylori* on asthma and allergy. *J. Asthma Allergy.* 2010;3:139–147. doi:10.2147/JAA.S8971; Oertli M, Müller A. *Helicobacter pylori* targets dendritic

cells to induce immune tolerance, promote persistence and confer protection against allergic asthma. *Gut Microbes.* 2012;3(6):566–571; Oertli M, Sundquist M, Hitzler I, et al. DC-derived Il-18 drives treg differentiation, murine *Helicobacter pylori*-specific immune tolerance, and asthma protection. *J Clin Invest.* 2012;122(3):1082–1096; Arnold IC, Dehzad N, Reuter S, et al. *Helicobacter pylori* infection prevents allergic asthma in mouse models through the induction of regulatory T cells. *J Clin Invest.* 2011;121(8):3088–3093; Arnold IC, Hitzler I, Müller A. The immunomodulatory properties of *Helicobacter pylori* confer protection against allergic and chronic inflammatory disorders. *Front Cell Infect Microbiol.* 2012;16(2):10. doi:10.3389/fcimb.2012.00010; Engler DB, Loenardi I, Hartung ML, et al. *Helicobacter pylori*-specific protection against inflammatory bowel disease requires the NLRP3 inflammasome and IL-18. *Inflamm Bowel Dis.* 2015;21(4):854–861.

119 **way to produce disease** Heller F, Fuss IJ, Nieuwenhuis EE, et al. Oxazolone colitis, a Th2 colitis model resembling ulcerative colitis, is mediated by IL-13-producing NK-T cells. *Immunity.* 2002;17(5):629–638.

119 **human ulcerative colitis** Fuss IJ, Strober W. The role of IL-13 and NK T cells in experimental and human ulcerative colitis. *Mucosal Immunol.* 2008;1(Suppl 1):S31–S3.

120 **avoid later-life disease** An D, Oh SF, Olszak T, et al. Sphingolipids from a symbiotic microbe regulate homeostasis of host intestinal natural killer T cells. *Cell.* 2014; 156(1–2):123–133; Olszak T, An D, Zeissig S, et al. Microbial exposure during early life has persistent effects on natural killer T cell function. *Science.* 2012;336(6080):489–493; Erturk-Hasdemir D, Kasper DL. Resident commensals shaping immunity. *Curr Opin Immunol.* 2013;25(4):450–455.

121 **approximately 675,000 in the United States** The pandemic: influenza strikes. *United States Department of Health and Human Services.* http://www .flu.gov/pandemic/history/1918/the_pandemic/influenza/. Accessed September 7, 2015.

121 **aberrant inflammatory response** The pandemic: influenza strikes. *United States Department of Health and Human Services.* http://www.flu.gov/ pandemic/history/1918/the_pandemic/influenza. Accessed July 18, 2015; Morens DM, Fauci AS. The 1918 influenza pandemic: insights for the 21st century. *J Infect Dis.* 2007;195(7):1018–1028. doi:10.1086/511989.

121 **among healthy young adults** Morens DM, Fauci AS. The 1918 influenza pandemic: insights for the 21st century. *J Infect Dis.* 2007;195(7):1018–1028; Morens DM, Taubenberger JK, Harvey HA, Memoli MJ. The 1918 influenza pandemic: lessons for 2009 and the future. *Crit Care Med.* 2010;38(Suppl 4):e10–e20.

122 ***Fatal Sequence: The Killer Within*** Tracey KJ. *Fatal Sequence: The Killer Within.* Chicago, IL: University of Chicago Press; 2005.

8. Patterns of Disease

125 **population were overweight** Obesity and overweight. *Centers for Disease Control and Prevention.* http://www.cdc.gov/nchs/fastats/obesity-overweight .htm. Accessed September 9, 2015.

125 **one-third of the population** NCHS Health E-Stat. Prevalence of overweight, obesity and extreme obesity among adults: United States, trends 1960–62 through 2005–2006. *Centers for Disease Control and Prevention.* http://www

.cdc.gov/nchs/data/hestat/overweight/overweight_adult.htm. Accessed September 9, 2015.

125 **children were obese** Obesity and overweight. *Centers for Disease Control and Prevention.* http://www.cdc.gov/nchs/fastats/obesity-overweight.htm. Accessed September 9, 2015.

125 **pro-inflammatory condition** Scarpellini E, Tack J. Obesity and metabolic syndrome: an inflammatory condition. *Dig Dis.* 2012;30(2):148–153; Olson S. Obesity's link to inflammation may unlock a switch responsible for bad fat. *Med Daily.* February 19, 2015. http://www.medicaldaily.com/obesitys-link -inflammation-may-unlock-switch-responsible-bad-fat-322692. Accessed September 9, 2015.

125 **diseases like cancer** Howe LR, Subbaramaiah K, Hudis CA, Dannenberg A J. Molecular pathways: adipose inflammation as a mediator of obesity-associated cancer. *Clin Cancer Res.* 2013;19(22):6074–6083.

125 **different types of cancer** Schauer P, Chand B. Reducing obesity-related co-morbidities through bariatric surgery. *Bariatrics Today.* 2006;3. https://my .clevelandclinic.org/ccf/media/files/Bariatric_Surgery/rccoe.pdf. Accessed September 10, 2015; Mayo Clinic Staff. Complications. Diseases and conditions: obesity. *Mayo Clinic.* http://www.mayoclinic.org/diseases-conditions/obesity /basics/complications/con-20014834. Accessed September 10, 2015; Lalwani AK, Katz K, Liu YH, et al. Obesity is associated with sensorineural hearing loss in adolescents. *Laryngoscope.* 2013;123(12):3178–3184; Kumar S, Han J, Li T, Qureshi AA. Obesity, waist circumference, weight change and the risk of psoriasis in US women. *J Eur Acad Dermatol Venereol.* 2013;27(10):1293–1298; Chuang YF, An Y, Bilgel M, et al. Midlife adiposity predicts earlier onset of Alzheimer's dementia, neuropathology and presymptomatic cerebral amyloid accumulation. *Mol Psychiatry.* 2015; Pauli-Pott U, Neidhard J, Heinzel-Gutenbrunner M, Becker K. On the link between attention deficit/ hyperactivity disorder and obesity: do comorbid oppositional defiant and conduct disorder matter? *Eur Child Adolesc Psychiatry.* 2014;23(7):531–537; Qin B, Yang M, Fu H, et al. Body mass index and the risk of rheumatoid arthritis: a systematic review and dose-response meta-analysis. *Arthritis Res Ther.* 2015;17(1):86; Gianfrancesco MA, Acuna B, Shen L, et al. Obesity during childhood and adolescence increases susceptibility to multiple sclerosis after accounting for established genetic and environmental risk factors. *Obes Res Clin Pract.* 2014;8(5):e435–e447; Behrens G, Matthews CE, Moore SC, et al. Body size and physical activity in relation to incidence of chronic obstructive pulmonary disease. *CMAJ.* 2014;186(12):e457–e469; Hall ME, do Carmo JM, da Silva AA, et al. Obesity, hypertension, and chronic kidney disease. *Int J Nephrol Renovasc Dis.* 2014;7:75–88. doi:10.2147/IJNRD.S39739.

128 **buildup can occur slowly** Hilgendorf I, Swirski FK, Robbins CS. Monocyte fate in atherosclerosis. *Arterioscler Thromb Vasc Biol.* 2015;35(2):272–279.

128 **infant decades earlier** Hoffman M. Atherosclerosis: your arteries age by age. Hardening of the arteries starts earlier than you may think. Heart Disease Health Center. *WebMD.* http://www.webmd.com/heart-disease/features/ atherosclerosis-your-arteries-age-by-age. Accessed July 25, 2015; Kelishadi R. Inflammation-induced atherosclerosis as a target for prevention of cardiovascular diseases from early life. *Open Cardiovasc Med J.* 2010;4:24–29. doi: 10.2174/1874192401004020024.

128 atherosclerosis will occur Kelishadi R. Inflammation-induced atherosclerosis as a target for prevention of cardiovascular diseases from early life. *Open Cardiovasc Med J.* 2010;4:24–29. doi:10.2174/1874192401004020024; Labayen I, Ortega FB, Sjöström M, Ruiz JR. Early life origins of low-grade inflammation and atherosclerosis risk in children and adolescents. *J Pediatr.* 2009;155(5):673–677.

129 unhealthy inflammation Kataoka Y, Puri R, Nicholls SJ. Inflammation, plaque progression and vulnerability: evidence from intravascular ultrasound imaging. *Cardiovasc Diagn Ther.* 2015;5(4):280–289.

129 risk of atherosclerosis Chistiakov DA, Bobryshev YV, Kozarov E, et al. Role of gut microbiota in the modulation of atherosclerosis-associated immune response. *Front Microbiol.* 2015;6:671. doi:10.3389/fmicb.2015.00671.

129 lose tissue function Ogura S, Shimosawa T. Oxidative stress and organ damages. *Curr Hypertens Rep.* 2014;16(8):452.

129 cells become cancerous Chew SH, Toyokuni S. Malignant mesothelioma as an oxidative stress-induced cancer: an update. *Free Radic Biol Med.* 2015;86:166–178.

130 strep throat Tan SY, Dee MK. Elie Metchnikoff (1845–1916): discoverer of phagocytosis. *Singapore Med J.* 2009;50(5):456–457. http://www.pasteur.fr/infosci/biblio/ressources/histoire/textes_integraux/metchnikoff/smjmetabi02009tan.pdf. Accessed September 8, 2015.

132 Alzheimer's disease, and diabetes Leading causes of death. *Centers for Disease Control and Prevention.* http://www.cdc.gov/nchs/fastats/leading-causes-of-death.htm. Accessed July 25, 2015.

133 prescription medication Stone K. The most prescribed medications by drug class. *About.com.* http://pharma.about.com/od/Sales_and_Marketing/a/The-Most-Prescribed-Medications-By-Drug-Class.htm. Accessed July 25, 2015.

134 being overweight Dietert RR, DeWitt JC, Germolec DR, Zelikoff JT. Breaking patterns of environmentally influenced disease for health risk reduction: immune perspectives. *Environ Health Perspect.* 2010;118(8):1091–1099. doi:10.1289/ehp.1001971.

134 other autoimmune conditions Kollipara S. Comorbidities associated with type I Diabetes. *NASN School Nurse.* 2010;25(1)19–21; Franzese A, Mozzillo E, Nugnes R, et al. Type 1 diabetes mellitus and co-morbidities. In: Wagner D, ed. *Type 1 Diabetes Complications. InTech*; 2011. doi:10.5772/24457. Available from: http://www.intechopen.com/books/type-1-diabetes-complications/type-1-diabetes-mellitus-and-co-morbidities. Accessed July 24, 2015; Farsani SF, Souverein PC, van der Vorst MMJ, et al. Chronic comorbidities in children with type I diabetes: a population-based cohort study. *Arch Dis Child.* 2015;100(8):763–768. doi:10.1136/archdischild-2014-307654; Butwicka A, Frisén L, Almqvist C, et al. Risks of psychiatric disorders and suicide attempts in children and adolescents with type I diabetes: a population-based cohort study. *Diabetes Care.* 2015;38(3):453–459.

134 associations with type 1 diabetes Harding JL, Shaw JE, Peeters A, et al. Cancer risk among people with type I and type 2 diabetes: disentangling true associations, detection bias, and reverse causation. *Diabetes Care.* 2015;38(2):264–270.

134 and type 1 diabetes Lauret E, Rodrigo L. Celiac disease and autoimmune-associated conditions. *Biomed Res Int.* 2013. Article ID 127589. doi:10.1155/2013/127589.

135 **small bowel adenocarcinoma** Dietert RR. Inflammatory bowel disease and celiac disease: environmental risks factors and consequences. In: Dietert RR, Luebke RW, eds. *Immunotoxicity, Immune Dysfunction, and Chronic Disease.* New York, NY: Humana Press, 2012:Chapter 12; Wagner G, Zeiler M, Berger G, et al. Eating disorders in adolescents with celiac disease: influence of personality characteristics and coping. *Eur Eat Disorders Rev.* 2015;23(5). doi:10.1002/erv.2376; Thompson JS, Lebwohl B, Reilly NR, et al. Increased incidence of eosinophilic esophagitis in children and adults with celiac disease. *J Clin Gastroenterol.* 2012;46(1):e6–e11.

135 **resistant to treatment** Blackmon K, Bluvstein J, MacAllister WS, et al. Treatment resistant epilepsy in autism spectrum disorder: increased risk for females. *Autism Res.* 2015. doi:10.1002/aur.1514.

135 **overall population** Kang V, Wagner GC, Ming X. Gastrointestinal dysfunction in children with autism spectrum disorders. *Autism Res.* 2014;7(4):501–506. doi:10.1002/aur.1386.

135 **ASD in children** Fadini CC, Lamônica DA, Fett-Conte AC, et al. Influence of sleep disorders on the behavior of individuals with autism spectrum disorder. *Front Hum Neurosci.* 2015;9:347. doi:10.3389/fnhum.2015.00347.

135 **majority of NCDs** Dietert RR, DeWitt JC, Germolec DR, Zelikoff JT. Breaking patterns of environmentally influenced disease for health risk reduction: immune perspectives. *Environ Health Perspect.* 2010;118(8):1091–1099. doi:10.1289/ehp.1001971.

135 **among ASD children** Theoharides TC. Autism spectrum disorders and mastocytosis. *Int J Immunopathol Pharmacol.* 2009;22(4):859–865.

135 **bipolar disorder** Croen LA, Zerbo O, Qian Y, et al. The health status of adults on the autism spectrum. *Autism.* 2015;19(7):814–823. doi:10.1177/1362361315557517.

136 **lead to depression** Bay-Richter C, Linderholm KR, Lim CK, et al. A role for inflammatory metabolites as modulators of the glutamate N-methyl-D-aspartate receptor in depression and suicidality. *Brain Behav Immun.* 2015;43:110–117; De Chant T. Depression may be caused by inflammation. *NOVA Next.* January 5, 2015. http://www.pbs.org/wgbh/nova/next/body/depression-may-caused-inflammation. Accessed September 9, 2015.

136 **much longer list** Vancampfort D, Mitchell AJ, De Hert M, et al. Type 2 diabetes in patients with major depressive disorder: a meta-analysis of prevalence estimates and predictors. *Depress Anxiety.* 2015;32(10):763–773; Carey M, Small H, Yoong SL, et al. Prevalence of comorbid depression and obesity in general practice: a cross-sectional survey. *Br J Gen Pract.* 2014;64(620):e122–e127; Dietert RR, DeWitt JC, Luebke RW. Reducing the prevalence of immune-based chronic disease. In: Dietert RR, Luebke RW. eds. *Immunotoxicity, Immune Dysfunction, and Chronic Disease*, New York, NY: Humana Press, 2012:Chapter 17.

136 **an all-time high** Wehrwein P. Astounding increase in antidepressant use by Americans. *Harvard Health Publications.* Harvard Medical School. October 20, 2011. http://www.health.harvard.edu/blog/astounding-increase-in-antidepressant-use-by-americans-201110203624. Accessed July 29, 2015.

137 **associated with them** Machiels K, Joossens M, Sabino J, et al. A decrease of the butyrate-producing species *Roseburia hominis* and *Faecalibacterium prausnitzii* defines dysbiosis in patients with ulcerative colitis. *Gut.* 2014;63(8):1275–1283; Lourenço TG, Heller D, Silva-Boghossian CM, et al. Mi-

crobial signature profiles of periodontally healthy and diseased patients. *J Clin Periodontol.* 2014;41(11):1027–1036.

137 Alzheimer's disease Zhao Y, Lukiw WJ. Microbiome-generated amyloid and potential impact on amyloidogenesis in Alzheimer's disease (AD). *J Nat Sci.* 2015;1(7). pii:e138.

137 Asthma Huang YJ, Boushey HA. The microbiome in asthma. *J Allergy Clin Immunol.* 2015;135(1):25–30.

137 Autism Buie T. Potential etiologic factors of microbiome disruption in autism. *Clin Ther.* 2015;37(5):976–983.

137 Autoimmune hepatitis Lin R, Zhou L, Zhang J, Wang B. Abnormal intestinal permeability and microbiota in patients with autoimmune hepatitis. *Int J Clin Exp Pathol.* 2015;8(5):5153–5160.

137 Breast cancer Xuan C, Shamonki JM, Chung A, et al. Microbial dysbiosis is associated with human breast cancer. *PLoS ONE.* 2014;9(1):e83744.doi:10.1371/journal.pone.0083744.

137 Cardiovascular disease Org E, Mehrabian M, Lusis AJ. Unraveling the environmental and genetic interactions in atherosclerosis: central role of the gut microbiota. *Atherosclerosis.* 2015;241(2):387–399.

138 Celiac disease Verdu EF, Galipeau HJ, Jabri B. Novel players in coeliac disease pathogenesis: role of the gut microbiota. *Nat Rev Gastroenterol Hepatol.* 2015;12(9):497–506.

138 Chronic kidney disease Montemurno E, Cosola C, Dalfino G, et al. What would you like to eat, Mr CKD Microbiota? A mediterranean diet, please! *Kidney Blood Press Res.* 2014;39(2–3):114–123.

138 Chronic obstructive pulmonary disease (COPD) Sze MA, Dimitriu PA, Suzuki M, et al. Host response to the lung microbiome in chronic obstructive pulmonary disease. *Am J Respir Crit Care Med.* 2015;192(4):438–445.

138 Colon cancer Gao Z, Guo B, Gao R, et al. Microbiota disbiosis is associated with colorectal cancer. *Front Microbiol.* 2015;6:20. doi:10.3389/fmicb.2015.00020.

138 Crohn's disease Schaubeck M, Clavel T, Calasan J, et al. Dysbiotic gut microbiota causes transmissible Crohn's disease-like ileitis independent of failure in antimicrobial defence. *Gut.* 2015. pii: gutjnl-2015-309333.

138 Depression Jiang H, Ling Z, Zhang Y, et al. Altered fecal microbiota composition in patients with major depressive disorder. *Brain Behav Immun.* 2015;48:186–194.

138 Food allergies Stefka AT, Feehley T, Tripathi P, et al. Commensal bacteria protect against food allergen sensitization. *Proc Natl Acad Sci USA.* 2014;111(36):13145–13150.

138 Hypertension Jose PA, Raj D. Gut microbiota in hypertension. *Curr Opin Nephrol Hypertens.* 2015;24(5):403–409.

138 Laryngeal cancer Gong H, Shi Y, Zhou X, et al. Microbiota in the throat and risk factors for laryngeal carcinoma. *Appl Environ Microbiol.* 2014;80(23):7356–7363.

138 Lung cancer Hosgood HD 3rd, Sapkota AR, Rothman N, et al. The potential role of lung microbiota in lung cancer attributed to household coal burning exposures. *Environ Mol Mutagen.* 2014;55(8):643–651.

138 Lupus Van Praet JT, Donovan E, Vanassche I, et al. Commensal microbiota influence systemic autoimmune responses. *EMBO J.* 2015;34(4):466–474.

138 Nonalcoholic fatty liver disease Abdul-Hai A, Abdallah A, Malnick SD. Influence of gut bacteria on development and progression of non-alcoholic fatty liver disease. *World J Hepatol.* 2015;7(12):1679–1684. doi:10.4254/wjh.v7.i12.1679.

138 Obesity Tilg H, Adolph TE. Influence of the human intestinal microbiome on obesity and metabolic dysfunction. *Curr Opin Pediatr.* 2015;27(4):496–501.

138 Osteoporosis Ohlsson C, Sjögren K. Effects of the gut microbiota on bone mass. *Trends Endocrinol Metab.* 2015;26(2):69–74.

138 Parkinson's disease Keshavarzian A, Green SJ, Engen PA, et al. Colonic bacterial composition in Parkinson's disease. *Mov Disord.* 2015;30(10):1351–1360.

138 Periodontal disease Lourenço TG, Heller D, Silva-Boghossian CM, et al. Microbial signature profiles of periodontally healthy and diseased patients. *J Clin Periodontol.* 2014;41(11):1027–1036.

138 Prostate cancer Yu H, Meng H, Zhou F, et al. Urinary microbiota in patients with prostate cancer and benign prostatic hyperplasia. *Arch Med Sci.* 2015;11(2):385–394.

138 Psoriasis Scher JU, Ubeda C, Artacho A, et al. Decreased bacterial diversity characterizes the altered gut microbiota in patients with psoriatic arthritis, resembling dysbiosis in inflammatory bowel disease. *Arthritis Rheumatol.* 2015;67(1):128–139.

138 Respiratory allergies Melli LC, do Carmo-Rodrigues MS, Araújo-Filho H, et al. Intestinal microbiota and allergic diseases: a systematic review. *Allergol Immunopathol (Madr).* 2015. pii: S0301-0546(15)00059-2.

138 Rheumatoid arthritis Brusca SB, Abramson SB, Scher JU. Microbiome and mucosal inflammation as extra-articular triggers for rheumatoid arthritis and autoimmunity. *Curr Opin Rheumatol.* 2014;26(1):101–107.

138 Schizophrenia Severance EG, Gressitt KL, Stallings CR, et al. Discordant patterns of bacterial translocation markers and implications for innate immune imbalances in schizophrenia. *Schizophr Res.* 2013;148(1–3):130–137.

138 Sudden infant death syndrome (SIDS) Highet AR, Berry AM, Bettelheim KA, Goldwater PN. Gut microbiome in sudden infant death syndrome (SIDS) differs from that in healthy comparison babies and offers an explanation for the risk factor of prone position. *Int J Med Microbiol.* 2014;304(5–6):735–741.

138 Type 1 diabetes Alkanani AK, Hara N, Gottlieb PA, et al. Alterations in intestinal microbiota correlate with susceptibility to type 1 diabetes. *Diabetes.* 2015;64(10):3510–3520.

138 Type 2 diabetes Upadhyaya S, Banerjee G. Type 2 diabetes and gut microbiome: at the intersection of known and unknown. *Gut Microbes.* 2015;6(2):85–92.

138 Ulcerative colitis Lavelle A, Lennon G, O'Sullivan O, et al. Spatial variation of the colonic microbiota in patients with ulcerative colitis and control volunteers. *Gut.* 2015;64(10):1553–1561.

138 Urothelial cancer Xu W, Yang L, Lee P, et al. Mini-review: perspective of the microbiome in the pathogenesis of urothelial carcinoma. *Am J Clin Exp Urol.* 2014;2(1):57–61.

139 to become obese Duca FA, Sakar Y, Lepage P, et al. Replication of obesity and associated signaling pathways through transfer of microbiota from obese-prone rats. *Diabetes.* 2014;63(5):1624–1636.

139 complete microbiome is present Esposito D, Damsud T, Wilson M, et al. Black currant anthocyanins attenuate weight gain and improve glucose metabolism in diet-induced obese mice with intact, but not disrupted, gut microbiome. *J Agric Food Chem.* 2015;63(27):6172–6180.

139 **altering the microbiome** Kim BS, Song MY, Kim H. The anti-obesity effect of *Ephedra sinica* through modulation of gut microbiota in obese Korean women. *J Ethnopharmacol.* 2014;152(3):532–539.

139 **treating the condition** Gut bacteria may affect whether a statin drug lowers cholesterol. Science News. *Science Daily.* October 14, 2011. http://www.science daily.com/releases/2011/10/111013184815.htm. Accessed July 25, 2015; Yoo DH, Kim IS, Van Le TK, et al. Gut microbiota-mediated drug interactions between lovastatin and antibiotics. *Drug Metab Dispos.* 2014;42(9):1508–1513.

140 **digoxin metabolism is compromised** Haiser HJ, Seim KL, Balskus EP, Turn-baugh PJ. Mechanistic insight into digoxin inactivation by *Eggerthella lenta* augments our understanding of its pharmacokinetics. *Gut Microbes.* 2014;5(2):233–238.

140 **cyclophosphamide chemotherapy** Iida N, Dzutsev A, Stewart CA, et al. Commensal bacteria control cancer response to therapy by modulating the tumor microenvironment. *Science.* 2013;342(6161):967–970; Viaud S, Saccheri F, Mignot G, et al. The intestinal microbiota modulates the anticancer immune effects of cyclophosphamide. *Science.* 2013;342(6161):971–976; Karin M, Jobin C, Balkwill F. Chemotherapy, immunity and microbiota—a new triumvirate? *Nat Med.* 2014;20(2):126–127.

9. The Six Causes of the Epidemic

143 **US infants and children** Chai G, Governale L, McMahon AW, et al. Trends of outpatient prescription drug utilization in US children, 2002–2010. *Pediatrics.* 2012;130(1):23–31.

143 **tissues and organs** Vangay P, Ward T, Gerber JS, Knights D. Antibiotics, pe-diatric dysbiosis, and disease. *Cell Host Microbe.* 2015;17(5):553–564; Saari A, Virta LJ, Sankilampi U, et al. Antibiotic exposure in infancy and risk of being overweight in the first 24 months of life. *Pediatrics.* 2015;135(4):617–626.

143 **complications often occurs** Dietert RR, Dietert JM. The microbiome and sustainable healthcare. *Healthcare.* 2015;3(1):100–129. doi:10.3390/health-care3010100; Llor C, Bjerrum L. Antimicrobial resistance: risk associated with antibiotic overuse and initiatives to reduce the problem. *Ther Adv Drug Saf.* 2014;5(6):229–241; Venekamp RP, Sanders S, Glasziou PP, et al. Antibiotics for acute otitis media in children. *Cochrane Database Syst Rev.* 2013;1:CD000219. doi:10.1002/14651858.CD000219.pub3; Milligan S, McCrery S. Should children with acute otitis media routinely be treated with antibiotics? No: most chil-dren older than two years do not require antibiotics. *Am Fam Physician.* 2013;88(7). Online. Accessed August 5, 2015.

144 **deaths each year** Antibiotic resistance threats in the United States, 2013. *Cen-ters for Disease Control and Prevention.* http://www.cdc.gov/drugresistance /threat-report-2013/. Accessed September 10, 2015.

144 **susceptibility to serious infections** Vangay P, Ward T, Gerber JS, Knights D. Antibiotics, pediatric dysbiosis, and disease. *Cell Host Microbe.* 2015;17(5):553–564.

146 **late 1960s in the UK** Dibner JJ, Richards JD. Antibiotic growth promoters in agriculture: history and mode of action. *Poult Sci.* 2005;84(4):634–643.

146 **via egg injections** Strom S. Antibiotics eliminated in hatchery, Perdue says. *The New York Times.* September 3, 2014. http://www.nytimes.com/2014/09/04/

business/perdue-eliminates-antibiotic-use-in-its-hatcheries.html. Accessed September 9, 2015.

146 **among major producers** Grow B, Huffstutter PJ. Major poultry farms routinely feed antibiotics to chickens. *Huff Post Green.* September 15, 2014. http://www.huffingtonpost.com/2014/09/15/poultry-farms-antibiotics-chickens_n_5822438.html. Accessed September 9, 2015.

146 **pathogen load during production** Dietert RR, Golemboski KA, Austic RE. Environment-immune interactions. *Poult Sci.* 1994;73(7):1062–1076.

146 **other media sources** Spotts PN. Controlling bacteria on the farm. *The Christian Science Monitor.* June 25, 1998. http://www.csmonitor.com/1998/0625/062598.feat.feat.7.html. Accessed September 10, 2015.

152 **such as bifidobacteria** LeBlanc JG, Milani C, de Giori GS, et al. Bacteria as vitamin suppliers to their host: a gut microbiota perspective. *Curr Opin Biotechnol.* 2013;24(2):160–168.

152 **westernized diet** Sonnenburg ED, Sonnenburg JL. Starving our microbial self: the deleterious consequences of a diet deficient in microbiota-accessible carbohydrates. *Cell Metab.* 2014;20(5):779–786.

153 **different than in the past** Cyril S, Oldroyd JC, Renzaho A. Urbanisation, urbanicity, and health: a systematic review of the reliability and validity of urbanicity scales. *BMC Public Health.* 2013;13:513. doi:10.1186/1471-2458-13-513.

153 **castle from the early 1100s** History of Edinburgh Castle. *Edinburgh Castle.* http://www.edinburghcastle.co.uk/history/. Accessed July 20, 2015.

153 **in the cramped spaces** Graham HG. *The Social Life of Scotland in the Eighteenth Century.* London, England: Adam and Charles Black; 1906.

154 **buildings as possible** Life and death in Old Edinburgh. *The Edinburgh Dungeon.* http://www.thedungeons.com/edinburgh/en/downloads/edinburgh_schools_resource_life_and_death.pdf. Accessed July 21, 2015.

154 **well-respected merchants** Fortescue WI. James Ker, member of parliament for Edinburgh, 1747–1754. *The Book of the Old Edinburgh Club.* 2014;10:17–44.

154 **"unhealthy hovel"** Chambers R, Thomson T. *A Biographical Dictionary of Eminent Scotsmen* (volume 5). Glasgow, Scotland: Blackie; 1854.

154 **healthier housing** The Scots Peerage. Paul JB, ed. *The Genealogist* (volume 2). Exeter, England: William Pollard & Company; 1878; Thomson T. *A Biographical Dictionary of Eminent Scotsmen* (volume 5). Glasgow, Scotland: Blackie; 1854.

154 **at the time** The Scots Peerage. Paul JB, ed. *The Genealogist* (volume 2) Exeter, England. William Pollard & Company; 1878; Chambers R. *Lives of Illustrious and Distinguished Scotsman, Forming a Complete Scottish Biographical Dictionary* (volume 3). Glasgow, Scotland: Blackie and Son; 1839.

155 **half of all deaths** City of Edinburgh council area—demographic factsheet. *National Records of Scotland.* http://www.nrscotland.gov.uk/files/statistics/council-area-data-sheets/city-of-edinburgh-factsheet.pdf. Accessed July 20, 2015.

155 **ten million people** World's population increasingly urban with more than half living in urban areas. *United Nations Department of Economic and Social Affairs.* July 10, 2014. http://www.un.org/en/development/desa/en/news/population/world-urbanization-prospects.html. Accessed July 20, 2015.

156 **most common killer, NCDs** Giles-Corti B, Badland H, Mavoa S, et al. Reconnecting urban planning with health: a protocol for the development and validation of national liveability indicators associated with noncommunicable

disease risk behaviours and health outcomes. *Public Health Res Pract.* 2014;25(1). pii: e2511405. doi:10.17061/phrp2511405; Cooper R, Boyko CT, Cooper C. Design for health: the relationship between design and noncommunicable diseases. *J Health Commun.* 2011;16(Suppl 2):134–157; Smit W, de Lannoy A, Dover RV, et al. Making unhealthy places: The built environment and noncommunicable diseases in Khayelitsha, Cape Town. *Health Place.* 2015;35:11–18.

156 contributed heavily to the deaths Geronimus AT. To mitigate, resist, or undo: addressing structural influences on the health of urban populations. In: Hynes HP, Lopez R, eds. *Urban Health: Readings in the Social, Built, and Physical Environments of U.S. Cities.* Sudbury, MA: Jones & Bartlett; 2009.

156 elevated systemic inflammation Viehmann A, Hertel S, Fuks K, et al. Long-term residential exposure to urban air pollution, and repeated measures of systemic blood markers of inflammation and coagulation. *Occup Environ Med.* 2015;72(9):656–663.

156 heart disease and asthma Ghosh R, Lurmann F, Perez L, et al. Near-roadway air pollution and coronary heart disease: burden of disease and potential impact of a greenhouse gas reduction strategy in Southern California. *Environ Health Perspect.* 2015. doi:10.1289/ehp.1408865; Cesaroni G, Badaloni C, Gariazzo C, et al. Long-term exposure to urban air pollution and mortality in a cohort of more than a million adults in Rome. *Environ Health Perspect.* 2013;121(3):324–331; Perez L, Declercq C, Iñiguez C, et al. Chronic burden of near-roadway traffic pollution in 10 European cities (APHEKOM network). *Eur Respir J.* 2013;42(3):594–560; Matsui EC. Environmental exposures and asthma morbidity in children living in urban neighborhoods. *Allergy.* 2014;69(5):553–558.

156 air pollution of cities Leon Hsu HH, Mathilda Chiu YH, Coull BA, et al. Prenatal particulate air pollution and asthma onset in urban children. Identifying sensitive windows and sex differences. *Am J Respir Crit Care Med.* 2015;192(9):1052–1059.

156 increases the risk of obesity Jerrett M, McConnell R, Wolch J, et al. Traffic-related air pollution and obesity formation in children: a longitudinal, multi-level analysis. *Environ Health.* 2014;13:49. doi:10.1186/1476-069X-1349; Ponticiello BG, Capozzella A, Di Giorgio V, et al. Overweight and urban pollution: preliminary results. *Sci Total Environ.* 2015;518–519:61–64.

157 Canada or Australia Weller C. This Chinese mega-city has more people than Canada or Australia. *Business Insider Australia.* July 9, 2015. http://www.businessinsider.com.au/chinese-mega-city-has-more-people-than-canada-argentina-or-australia-2015-7. Accessed July 20, 2015.

157 130 million people Johnson I. Yanjiao: ambitious supercity of 130 million people developing around Beijing. *The Economic Times.* July 20, 2015. http://economictimes.indiatimes.com/news/international/business/yanjiao-ambitious-supercity-of-130-million-people-developing-around-beijing/articleshow/48142531.cms. Accessed July 20, 2015.

157 hygiene hypothesis Von Mutius E. 99th Dahlem conference on infection, inflammation and chronic inflammatory disorders: farm lifestyles and the hygiene hypothesis. *Clin Exp Immunol.* 2010;160(1):130–135.

157 risk of multiple NCDs Riedler J, Braun-Fahrländer C, Eder W, et al. Exposure to farming in early life and development of asthma and allergy: a cross-sectional survey. *Lancet.* 2001;358(9288):1129–1133; Ege MJ, Herzum I, Büchele G, et al. Prenatal exposure to a farm environment modifies atopic sensitization at birth. *J Allergy Clin Immunol.* 2008;122(2):407–412; Nicolaou N, Siddique N, Custovic A. Allergic disease in urban and rural populations: increasing prevalence with increasing urbanization. *Allergy.* 2005;60(11):1357–1360; Growing up on a farm directly affects regulation of the immune system, study finds. Science News. *Science Daily.* February 8, 2012. http://www.sciencedaily.com/releases/2012/02/120208132549.htm. Accessed September 10, 2015; Growing up on livestock farm halves risk of inflammatory bowel diseases. Science News. *Science Daily.* July 11, 2014. http://www.sciencedaily.com/releases/2014/07/140711101347.htm. Accessed September 10, 2015.

157 **movie version are planned** Busch A. "Green Acres" moving from Hooterville to Hollywood. Feature film. Broadway play in the works. *Deadline Hollywood.* May 2, 2014. http://deadline.com/2014/05/green-acres-movie-broadway -show-722481/. Accessed July 23, 2015.

157 **allergic disease and asthma** Von Ehrenstein OS, Von Mutius E, Illi S, et al. Reduced risk of hay fever and asthma among children of farmers. *Clin Exp Allergy.* 2000;30(2):187–193; Riedler J, Braun-Fahrländer C, Eder W, et al. Exposure to farming in early life and development of asthma and allergy: a cross-sectional survey. *Lancet.* 2001;358(9288):1129–1133; Von Mutius E, Radon K. Living on a farm: impact on asthma induction and clinical course. *Immunol Allergy Clin North Am.* 2008;28(3):631–647.

157–58 **affect immune development** Schröder PC, Li J, Wong GW, Schaub B. The rural-urban enigma of allergy: what can we learn from studies around the world? *Pediatr Allergy Immunol.* 2015;26(2):95–102. doi:10.1111/pai.12341; Lluis A, Depner M, Gaugler B, et al. Increased regulatory T-cell numbers are associated with farm milk exposure and lower atopic sensitization and asthma in childhood. *J Allergy Clin Immunol.* 2014;133(2):551–559; Pfefferle PI, Büchele G, Blümer N, et al. Cord blood cytokines are modulated by maternal farming activities and consumption of farm dairy products during pregnancy: the PASTURE Study. *J Allergy Clin Immunol.* 2010;125(1):108–115.

158 **would be produced** Schuijs MJ, Willart MA, Vergote K, et al. Farm dust and endotoxin protect against allergy through A20 induction in lung epithelial cells. *Science.* 2015;349(6252):1106–1110.

158 **noncommunicable diseases** Thysen AH, Larsen JM, Rasmussen MA, et al. Prelabor cesarean section bypasses natural immune cell maturation. *J Allergy Clin Immunol.* 2015;136(4):1123–1125; Mesquita DN, Barbieri MA, Goldani HA, et al. Cesarean section is associated with increased peripheral and central adiposity in young adulthood: cohort study. *PLoS ONE.* 2013;8(6):e66827. doi:10.1371/journal.pone.0066827; Jakobsson HE, Abrahamsson TR, Jenmalm MC, et al. Decreased gut microbiota diversity, delayed Bacteroidetes colonisation and reduced Th1 responses in infants delivered by caesarean section. *Gut,* 2014;63(4):559–566; Dietert RR. Natural childbirth and breastfeeding as preventive measures of immune-microbiome dysbiosis and misregulated inflammation. *J Anc Dis Prev Rem.* 2013;1(2):103. http://dx.doi.org/10.4172/jadpr1000103.

158 **complementary therapy is used** Miniello VL, Colasanto A, Cristofori F, et al. Gut microbiota biomodulators, when the stork comes by the scalpel. *Clin Chim Acta.* 2015;451(Pt A):88–96. doi:10.1016/j.cca.2015.01.022.

159 **global cesarean deliveries** Dietert RR. Natural childbirth and breastfeeding as preventive measures of immune-microbiome dysbiosis and misregulated inflammation. *J Anc Dis Prev Rem.* 2013;1(2):103. doi:10.4172.jadpr1000103.

159 dying women who were with child Cesarean section—a brief history, part I. *US National Library of Medicine.* https://www.nlm.nih.gov/exhibition/cesarean/part1.html. Accessed July 22, 2015.

159 until the 1580s Cesarean section—a brief history, part I. *US National Library of Medicine.* https://www.nlm.nih.gov/exhibition/cesarean/part1.html. Accessed July 22, 2015.

159 between 1996 and 2007 Recent trends in cesarean delivery in the United States. NCHS Data Brief. *Centers for Disease Control and Prevention.* http://www.cdc.gov/nchs/data/databriefs/db35.htm. Accessed July 22, 2015.

159 states and ethnic groups Recent trends in cesarean delivery in the United States. NCHS Data Brief. *Centers for Disease Control and Prevention.* http://www.cdc.gov/nchs/data/databriefs/db35.htm. Accessed July 22, 2015.

159 years 1997 and 2006 Karlström A, Lindgren H, Hildingsson I. Maternal and infant outcome after caesarean section without recorded medical indication: findings from a Swedish case-control study. *BJOG.* 2013;120(4):479–486.

159 between 1990 and 2008 Bragg F, Cromwell DA, Edozien LC, et al. Variation in rates of caesarean section among English NHS trusts after accounting for maternal and clinical risk: cross sectional study. *BMJ.* 2010:341(7777):c5065. doi:10.1136/bmj.c5065.

159 England, 24 percent Bragg F, Cromwell DA, Edozien LC, et al. Variation in rates of caesarean section among English NHS trusts after accounting for maternal and clinical risk: cross sectional study. *BMJ.* 2010;341(7777):c5065.

159 US, 33 percent Births—method of delivery. FastStats. *Centers for Disease Control and Prevention.* http://www.cdc.gov/nchs/fastats/delivery.htm. Accessed July 22, 2015.

159 India, 40 percent Ajeet S, Nandkishore K. The boom in unnecessary caesarean surgeries is jeopardizing women's health. *Health Care Women Int.* 2013;34(6):513–521.

159 country of origin Teixeira C, Correia S, Victora CG, Barros H. The Brazilian preference: cesarean delivery among immigrants in Portugal. *PLoS ONE.* 2013;8(3):e60168. doi:10.1371/journal.pone.0060168.

159 China, 46 percent Lumbiganon P, Laopaiboon M, Gülmezoglu AM, et al. Method of delivery and pregnancy outcomes in Asia: the WHO global survey on maternal and perinatal health 2007–08. *Lancet.* 2010;375(9713):490–499.

161 first year of life Bäckhed F, Roswall J, Peng Y, et al. Dynamics and stabilization of the human gut microbiome during the first year of life. *Cell Host Microbe.* 2015;17(5):690–703.

162 "Moms Deliver Twice!" Frese SA, Mills DA. Birth of the infant gut microbiome: moms deliver twice! *Cell Host Microbe.* 2015;17(5):543–544.

162 birth by C-section Mastromarino P, Capobianco D, Miccheli A, et al. Administration of a multistrain probiotic product (VSL#3) to women in the perinatal period differentially affects breast milk beneficial microbiota in relation to mode of delivery. *Pharmacol Res.* 2015;95–96:63–70.

163 further maturation after birth Dietert RR, Dietert J. *Strategies for Protecting Your Child's Immune System.* Hackensack, NJ: World Scientific Publishing; 2010.

163 suppressed in the infant Jakobsson HE, Abrahamsson TR, Jenmalm MC, et al. Decreased gut microbiota diversity, delayed Bacteroidetes colonisation and reduced Th1 responses in infants delivered by caesarean section. *Gut.* 2014;63(4):559–566.

163 **Th1 immune responses** Puff R, D'Orlando O, Heninger AK, et al. Compromised immune response in infants at risk for type 1 diabetes born by caesarean section. *Clin Immunol.* 2015;160(2):282–285.

163 **vaginally delivered children** Van Berkel AC, den Dekker HT, Jaddoe VW, et al. Mode of delivery and childhood fractional exhaled nitric oxide, interrupter resistance and asthma: the Generation R study. *Pediatr Allergy Immunol.* 2015;26(4):330–336.

163 **from vaginal births** Sevelsted A, Stokholm J, Bønnelykke K, Bisgaard H. Cesarean section and chronic immune disorders. *Pediatrics.* 2015;135(1):e92–e98.

164 **obesity** Kuhle S, Tong OS, Woolcott CG. Association between caesarean section and childhood obesity: a systematic review and meta-analysis. *Obes Rev.* 2015;16(4):295–303.

164 **autism spectrum disorders and ADHD** Curran EA, O'Neill SM, Cryan JF, et al. Research review: Birth by caesarean section and development of autism spectrum disorder and attention-deficit/hyperactivity disorder: a systematic review and meta-analysis. *J Child Psychol Psychiatry.* 2015;56(5):500–508.

164 **high blood pressure** Horta BL, Gigante DP, Lima RC, et al. Birth by caesarean section and prevalence of risk factors for non-communicable diseases in young adults: a birth cohort study. *PLoS ONE.* 2013;8(9):e74301. doi:10.1371/journal.pone.0074301.

164 **celiac disease** Mårild K, Stephansson O, Montgomery S, et al. Pregnancy outcome and risk of celiac disease in offspring: a nationwide case-control study. *Gastroenterology.* 2012;142(1):39–45.

164 **risk of food allergies** Koplin J, Allen K, Gurrin L, et al. Is caesarean delivery associated with sensitization to food allergens and IgE-mediated food allergy: a systematic review. *Pediatr Allergy Immunol.* 2008;19(8):682–687.

164 **atopic dermatitis** Lee SY, Yu J, Ahn KM, et al. Additive effect between IL-13 polymorphism and cesarean section delivery/prenatal antibiotics use on atopic dermatitis: a birth cohort study (COCOA). *PLoS ONE.* 2014;9(5):e96603.

165 **serious birth defects** Fintel B, Samaras TS, Carias E. The thalidomide tragedy: lessons for drug safety and regulation. *HELIX.* July 28, 2009. https://helix.northwestern.edu/article/thalidomide-tragedy-lessons-drug-safety-and-regulation. Accessed July 23, 2015.

165 **major health hazard** Asbestos health effects. *ATSDR.* http://www.atsdr.cdc.gov/asbestos/asbestos/health_effects/. Accessed July 23, 2015.

165 **exposed children** Lanphear BP, Hornung R, Khoury J, et al. Low-level environmental lead exposure and children's intellectual function: an international pooled analysis. *Environ Health Perspect.* 2005;113(7): 894-899; Wells EM, Bonfield TL, Dearborn DG, Jackson LW. The relationship of blood lead with immunoglobulin E, eosinophils, and asthma among children: NHANES 2005–2006. *Int J Hyg Environ Health.* 2014;217(2–3):196–204.

165 **numerous physiological systems** Menard S, Guzylack-Piriou L, Leveque M, et al. Food intolerance at adulthood after perinatal exposure to the endocrine disruptor bisphenol A. *FASEB J.* 2014;28(11):4893–4900; Strakovsky RS, Wang H, Engeseth NJ, et al. Developmental bisphenol A (BPA) exposure leads to sex-specific modification of hepatic gene expression and epigenome at birth that may exacerbate high-fat diet-induced hepatic steatosis. *Toxicol Appl Pharmacol.* 2015;284(2):101–112; Mileva G, Baker SL, Konkle ATM, Bielajew C.

Bisphenol-A: epigenetic reprogramming and effects on reproduction and behavior. *Int J Environ Res Public Health.* 2014;11(7):7537–7561.

165 **toxicity and cancer** Gross L. Flame retardants in consumer products are linked to health and cognitive problems. *The Washington Post.* April 15, 2013. https://www.washingtonpost.com/national/health-science/flame-retardants -in-consumer-products-are-linked-to-health-and-cognitive-problems/2013/04/ 15/f5c7b2aa-8b34-11e2-9838-d62f083ba93f_story.html. Accessed July 23, 2015.

165 *Legally Poisoned* Cranor CF. *Legally Poisoned: How the Law Puts Us at Risk from Toxicants.* Cambridge, MA: Harvard University Press; 2011.

167 **food emulsifiers** Chassaing B, Koren O, Goodrich JK, et al. Dietary emulsifiers impact the mouse gut microbiota promoting colitis and metabolic syndrome. *Nature.* 2015;519(7541):92–96.

167 **following some surgeries** Kastl KG, Betz CS, Siedek V, Leunig A. Effect of carboxymethylcellulose nasal packing on wound healing after functional endoscopic sinus surgery. *Am J Rhinol Allergy.* 2009;23(1):80–84.

168 **including *Lactobacillus* species** Clair E, Linn L, Travert C, et al. Effects of Roundup(®) and glyphosate on three food microorganisms: *Geotrichum candidum, Lactococcus lactis* subsp. cremoris and *Lactobacillus delbrueckii* subsp. bulgaricus. *Curr Microbiol.* 2012;64(5):486–491; Shehata AA, Kühnert M, Haufe S, Krüger M. Neutralization of the antimicrobial effect of glyphosate by humic acid in vitro. *Chemosphere.* 2014;104:258–261.

10. Precision Medicine Envisaged

172 **"Medicine's Next Frontier"** Tirrell M. Medicine's next frontier: the microbiome. CNBC. *Biotech and Pharmaceuticals.* June 19, 2015. http://www.cnbc .com/2015/06/19/medicines-next-frontier-the-microbiome.html. Accessed July 28, 2015.

173 **available hospital bugs** House L. Would you "seed" your baby? The surprising post-birth trend on the rise that sees mothers who have c-sections bathe newborns in vaginal fluid. *Mail Online.* June 24, 2015. http://www.dailymail .co.uk/femail/article-3136832/Would-seed-baby-alarming-post-birth-trend -rise-Australia-sees-mothers-C-sections-bathe-newborns-vaginal-fluid.html. Accessed September 10, 2015.

174 **diversity in human milk** Cabrera-Rubio R, Collado MC, Laitinen K, et al. The human milk microbiome changes over lactation and is shaped by maternal weight and mode of delivery. *Am J Clin Nutr.* 2012;96(3):544–551.

174 **neuroactive compounds** Wall R, Cryan JF, Ross RP, et al. Bacterial neuroactive compounds produced by psychobiotics. *Adv Exp Med Biol.* 2014;817:221– 239.

174 **the drug Prozac** Kantak PA, Bobrow DN, Nyby JG. Obsessive-compulsive-like behaviors in house mice are attenuated by a probiotic (*Lactobacillus rhamnosus* GG). *Behav Pharmacol.* 2014;25(1):71–79.

175 **pharmaceutical-based therapies** Mohammadi AA, Jazayeri S, Khosravi-Darani K, et al. The effects of probiotics on mental health and hypothalamic-pituitary-adrenal axis: a randomized, double-blind, placebo-controlled trial in petrochemical workers. *Nutr Neurosci.* 2015. doi:10.1179/1476830515Y .0000000023.

175 **these treatments work** Nelson MH, Diven MA, Huff LW, Paulos CM. Harnessing the microbiome to enhance cancer immunotherapy. *J Immunol Res.* 2015;2015:Article ID368736.

176 **production of butyrate** Ohashi Y, Sumitani K, Tokunaga M, et al. Consumption of partially hydrolysed guar gum stimulates Bifidobacteria and butyrate-producing bacteria in the human large intestine. *Benef Microbes.* 2015;6(4):451–455.

176 **food more efficiently** Pruszynska-Oszmalek E, Kolodziejski PA, Stadnicka K, et al. In ovo injection of prebiotics and synbiotics affects the digestive potency of the pancreas in growing chickens. *Poult Sci.* 2015;94(8):1909–1916.

177 **fight against NCDs** Sharon G, Garg N, Debelius J, et al. Specialized metabolites from the microbiome in health and disease. *Cell Metab.* 2014;20(5):719–730.

181 **fewer complications** Vicari E, La Vignera S, Castiglione R, et al. Chronic bacterial prostatitis and irritable bowel syndrome: effectiveness of treatment with rifaximin followed by the probiotic VSL#3. *Asian J Androl.* 2014;16(5):735–739.

182 **produce allergies** Giacomin P, Croese J, Krause L, et al. Suppression of inflammation by helminths: a role for the gut microbiota? *Philos Trans R Soc Lond B Biol Sci.* 2015;370(1675). pii: 20140296; Afifi MA, Jiman-Fatani AA, El Saadany S, Fouad MA. Parasites–allergy paradox: disease mediators or therapeutic modulators. *J Microscopy and Ultrastruct.* 2015;3(2):53–61.

182 **compared with controls** Dhiman RK, Rana B, Agrawal S, et al. Probiotic VSL#3 reduces liver disease severity and hospitalization in patients with cirrhosis: a randomized, controlled trial. *Gastroenterology.* 2014;147(6):1327–1337.

182 **autoimmune disease** Mardini HE, Grigorian AY. Probiotic mix VSL#3 is effective adjunctive therapy for mild to moderately active ulcerative colitis: a meta-analysis. *Inflamm Bowel Dis.* 2014;20(9):1562–1567.

182 **heart disease** Sanaie S, Ebrahimi-Mameghani M, Mahmoodpoor A, et al. Effect of a probiotic preparation (VSL#3) on cardiovascular risk parameters in critically-ill patients. *J Cardiovasc Thorac Res.* 2013;5(2):67–70.

182 **liver disease in children** Miccheli A, Capuani G, Marini F, et al. Urinary (1) H-NMR-based metabolic profiling of children with NAFLD undergoing VSL#3 treatment. *Int J Obes.* 2015;39(7):1118–1125.

182 **irritable bowel syndrome** Wong RK, Yang C, Song GH, et al. Melatonin regulation as a possible mechanism for probiotic (VSL#3) in irritable bowel syndrome: a randomized double-blinded placebo study. *Dig Dis Sci.* 2015;60(1):186–194.

183 **responses to food allergens** Cosenza L, Nocerino R, Di Scala C, et al. Bugs for atopy: the *Lactobacillus rhamnosus* GG strategy for food allergy prevention and treatment in children. *Benef Microbes.* 2015;6(2):225–232.

183 **3.6 percent of controls** Tang MK, Ponsonby AL, Orsini F, et al. Administration of a probiotic with peanut oral immunotherapy: a randomized trial. *J Allergy Clin Immunol.* 2015;135(3):737–744.

183 **supporting the arthritis** Vaghef-Mehrabany E, Alipour B, Homayouni-Rad A, et al. Probiotic supplementation improves inflammatory status in patients with rheumatoid arthritis. *Nutrition.* 2014;30(4):430–435.

183 **overweight/obese adults** Dao MC, Everard A, Aron-Wisnewsky J, et al. *Akkermansia muciniphila* and improved metabolic health during a dietary intervention in obesity: relationship with gut microbiome richness and ecology. *Gut.* June 22, 2015. pii: gutjnl-2014-308778.

184 **problematic microbes** Easton J. Fecal "transplant" helps one-year-old beat relentless infection. *Science Life*. January 2, 2013. http://sciencelife.uchospitals .edu/2013/01/02/fecal-transplant-helps-one-year-old-beat-relentless-infection/. Accessed July 28, 2015; Kahn SA, Young S, Rubin DT. Colonoscopic fecal microbiota transplant for recurrent *Clostridium difficile* infection in a child. *Am J Gastroenterol*. 2012;107(12):1930–1931.

184 **colonoscopy procedure** Patel NC, Griesbach CL, DiBaise JK, Orenstein R. Fecal microbiota transplant for recurrent *Clostridium difficile* infection: Mayo Clinic in Arizona experience. *Mayo Clin Proc*. 2013;88(8):799–805.

184 **90 percent cure rate** Orally delivered microbial therapy has potential beyond CDI. Clinical Updates. *Mayo Clinic*. http://www.mayoclinic.org/medical -professionals/clinical-updates/digestive-diseases/orally-delivered-microbial -therapy-has-potential-beyond-cdi. Accessed July 30, 2015.

184 **capsules taken orally** Youngster I, Russell GH, Pindar C, et al. Oral, capsulized, frozen fecal microbiota transplantation for relapsing *Clostridium difficile* infection. *JAMA*. 2014;312(17):1772–1778.

185 **the donated poop** Moayyedi P, Surette MG, Kim PT, et al. Fecal microbiota transplantation induces remission in patients with active ulcerative colitis in a randomized controlled trial. *Gastroenterology*. 2015;149(1):102–109; Hays B. Fecal transplants used successfully to treat ulcerative colitis. *UPI*. July 3, 2015. http:// www.upi.com/Health_News/2015/07/03/Fecal-transplants-used-successfully -to-treat-ulcerative-colitis/8371435935830/. Accessed July 20, 2015; Berman J. Inflammatory bowel disease "cured" with fecal transplant. *Voice of America*. January 22, 2015. http://www.voanews.com/content/inflammatory-bowel -disease-cured-with-fecal-transplant/2609711.html. Accessed July 30, 2015.

185 **multiple sclerosis** Surana NK, Kasper DL. The yin yang of bacterial polysaccharides: lessons learned from B. fragilis PSA. *Immunol Rev*. 2012;245(1):13–26.

186 **neurodegenerative conditions** MacFabe DF. Enteric short-chain fatty acids: microbial messengers of metabolism, mitochondria, and mind: implications in autism spectrum disorders. *Microb Ecol Health Dis*. 2015;26:28177; Valvassori SS, Varela RB, Arent CO, et al. Sodium butyrate functions as an antidepressant and improves cognition with enhanced neurotrophic expression in models of maternal deprivation and chronic mild stress. *Curr Neurovasc Res*. 2014;11(4):359–366; Sharma S, Taliyan R, Singh S. Beneficial effects of sodium butyrate in 6-OHDA induced neurotoxicity and behavioral abnormalities: modulation of histone deacetylase activity. *Behav Brain Res*. 2015;291:306–314.

186 **microbiome-based therapeutics** Fukami H, Tachimoto H, Kishi M, et al. Acetic acid bacterial lipids improve cognitive function in dementia model rats. *J Agric Food Chem*. 2010;58(7):4084-4089.

11. You, the Volatile Organic Compound

190 **chemical tribromoanisole** 2,4,6-Tribromoanisole (TBA) Programs. *Pharma Resource Group Inc*. http://www.pharmarg.com/tribromoanisole.html. Accessed November 13, 2015.

190 **certain amino acids** Brennan J. More than just "blue": the world's top 10 smelly cheeses. *Fine Dining Lovers*. October 2, 2012. https://www.finedining lovers.com/stories/blue-cheese-list-smelly-cheeses/. Accessed July 31, 2015; Sourabié AM, Spinnler HE, Bourdat-Deschamps M, et al. S-methyl thioesters

are produced from fatty acids and branched-chain amino acids by brevibacteria: focus on L-leucine catabolic pathway and identification of acyl-CoA intermediates. *Appl Microbiol Biotechnol.* 2012;93(4):1673–1683.

190 thousand different VOCs "Sniffer" to unlock secrets of chocolate. *The Press* (New Zealand). September 21, 2008. http://www.stuff.co.nz/the-press/637889/ Sniffer-to-unlock-secrets-of-chocolate. Accessed July 31, 2015.

191 aroma and taste Gupta C, Prakash D, Gupta S. A biotechnological approach to microbially based perfumes and flavours. *J Microbiol Exp.* 2015;2(1):00034. doi:10.15406/jmen.2015.01.00034.

191 can readily detect Estapé N. Perfume: why does it smell different on each person? *The Healthy Skin Blog.* January 17, 2013. http://www.thehealthyskin blog.org/perfume-why-does-it-smell-different-on-each-person/. Accessed August 1, 2015.

191 repair the epithelial barrier Geirnaert A, Steyaert A, Eeckhaut V, et al. *Butyricicoccus pullicaecorum*, a butyrate producer with probiotic potential, is intrinsically tolerant to stomach and small intestine conditions. *Anaerobe.* 2014;30:70-74; Eeckhaut V, Machiels K, Perrier C, et al. *Butyricicoccus pullicaecorum* in inflammatory bowel disease. *Gut.* 2013;62(12):1745–1752.

191 ten parts per million Butyric acid, a very smelly molecule. *The Chronicle Flask.* November 29, 2014. https://thechronicleflask.wordpress.com/2014/11/ 29/butyric-acid-a-very-smelly-molecule/. Accessed July 31, 2015.

192 pee-mails and nosebooks Gadbois S, Reeve C. Canine olfaction: scent, sign, and situation. In: Horowitz A, ed. *Domestic Dog Cognition and Behavior.* Berlin, Germany: Springer-Verlag; 2014:3–29.

192 life-threatening situation Frequently asked questions about medical assistance & diabetic alert dogs. *Dogs4Diabetics, Inc.* http://www.dogs4diabetics .com/about-us/faq/. Accessed July 30, 2015.

192 signature of the disease Sonoda H, Kohnoe S, Yamazato T, et al. Colorectal cancer screening with odour material by canine scent detection. *Gut.* 2011;60:814–819.

192 lung and breast cancers McCulloch M, Jezierski T, Broffman M, et al. Diagnostic accuracy of canine scent detection in early- and late-stage lung and breast cancers. *Integr Cancer Ther.* 2006;5(1):30–39.

192 ovarian cancers Horvath G, Järverud GA, Järverud S, Horváth I. Human ovarian carcinomas detected by specific odor. *Integr Cancer Ther.* 2008;7(2):76–80.

192 bladder cancer Willis CM, Britton LE, Harris R, et al. Volatile organic compounds as biomarkers of bladder cancer: sensitivity and specificity using trained sniffer dogs. *Cancer Biomark.* 2010–2011;8(3):145–153.

193 lower concentrations Butyric acid, a very smelly molecule. *The Chronicle Flask.* November 29, 2014. https://thechronicleflask.wordpress.com/2014/11/ 29/butyric-acid-a-very-smelly-molecule/. Accessed July 31, 2015.

193 mold, based on scent Kauhanen E, Harri M, Nevalainen A, Nevalainen T. Validity of detection of microbial growth in buildings by trained dogs. *Environ Int.* 2002;28(3):153–157.

193 instruments can detect it Bomers MK, van Agtmael MA, Luik H, et al. Using a dog's superior olfactory sensitivity to identify *Clostridium difficile* in stools and patients: proof of principle study. *BMJ.* 2012;345:e7396.

194 over long distances Chiacchia K. Search-and-rescue dogs function and deployment. *Allegheny Mountain Rescue Group.* http://www.amrg.info/canine

-sar/become-a-sar-dog-handler/16-canine-sar/canine-sar/24-search-and-rescue
-dogs-function-and-deployment. Accessed August 2, 2015.

194 **water-quality remediation** Van De Werfhorst LC, Murray JL, Reynolds S, et al. Canine scent detection and microbial source tracking of human waste contamination in storm drains. *Water Environ Res.* 2014;86(6):550–558.

194 **correct the problem** Van De Werfhorst LC, Murray JL, Reynolds S, et al. Canine scent detection and microbial source tracking of human waste contamination in storm drains. *Water Environ Res.* 2014;86(6):550–558; Jensen M. Bacteria sniffing canines help track sources of contamination in Kirkland's Juanita Creek and Seattle's Thornton Creek. *Kirkland Patch.* May 5, 2014. http://patch.com/washington/kirkland/bacteria-sniffing-canines-help-track-sources-of-contamination-in-kirklands-juanita-creek-and-seattles-thornton-creek. Accessed August 5, 2015.

194 **causes tuberculosis** Giant rats to sniff out tuberculosis. *New Scientist.* December 15, 2003. https://www.newscientist.com/article/dn4488-giant-rats-to-sniff-out-tuberculosis/. Accessed August 1, 2015.

194–95 **fart-focused papers** Swift N. The adventures of M.D. Levitt, M.D. *Annals of Improbable Res.* 2006;12(3):26.

195 **autism spectrum disorder** Singh K, Connors SL, Macklin EA, et al. Sulforaphane treatment of autism spectrum disorder (ASD). *Proc Nat Acad Sci.* 2014;111(43):15550–15555.

196 **completely odorless** Shelly WB, et al. Axillary odor: experimental study of the role of bacteria, apocrine sweat, and deodorants. *AMA Arch Derm Syphilol.* 1953;68(3):430–446.

196 **personal microbial cloud** Algar J. Your own personal germ cloud: how your microbes follow you around. *Tech Times.* August 29, 2014. http://www.techtimes.com/articles/14373/20140829/your-own-personal-germ-cloud-how-your-microbes-follow-you-around.htm. Accessed August 2, 2015; Lax S, Smith DP, Hampton-Marcell J, et al. Longitudinal analysis of microbial interaction between humans and the indoor environment. *Science.* 2014;345(6200):1048–1052.

196 **just other people** Campbell C, Gries R, Kashkin G, Gries G. Organosulphur constituents in garlic oil elicit antennal and behavioural responses from the yellow fever mosquito. *J Appl Entomol.* 2011;135(5):374–381.

196 **skin microbiome** Verhulst NO, Qiu YT, Beijleveld H, et al. Composition of human skin microbiota affects attractiveness to malaria mosquitoes. *PLoS ONE.* 2011;6(12):e28991. doi:10.1371/journal.pone.0028991.

197 **details of the skin microbiome** Bouslimani A, Porto C, Rath CM, et al. Molecular cartography of the human skin surface in 3D. *Proc Natl Acad Sci USA.* 2015;112(17):e2120–e2129.

197 **our co-partners** Bouslimani A, Porto C, Rath CM, et al. Molecular cartography of the human skin surface in 3D. *Proc Natl Acad Sci USA.* 2015;112(17):e2120–e2129.

197 **oils on the skin** Cooks RG, Jarmusch AK, Ferreira CR, Pirro V. Skin molecule maps using mass spectrometry. *Proc Natl Acad Sci USA.* 2015;112(17):5261–5262.

198 **natural deodorants** Troccaz M, Gaïa N, Beccucci S, et al. Mapping axillary microbiota responsible for body odours using a culture-independent approach. *Microbiome.* 2015;3(1):3. doi:10.1186/S40168-014-0064-3.

198 **artificial sweat** Callewaert C, Buysschaert B, Vossen E, et al. Artificial sweat

composition to grow and sustain a mixed human axillary microbiome. *J Microbiol Methods.* 2014;103:6–8.

198 environment of your mouth Zaura E, Mira A. Editorial: the oral microbiome in an ecological perspective. *Front Cell Infect Microbiol.* 2015;5:39. doi:10.3389/fcimb.2015.00039.

199 caused by your oral microbes Oral chroma: aiming to set the standard in halitosis measuring devices. *Abimedical.* http://www.abimedical.com/en/medical/product_01.html. Accessed August 1, 2015.

199 white wine grapes Muñoz-González C, Cueva C, Ángeles Pozo-Bayón M, Moreno-Arribas MV. Ability of human oral microbiota to produce wine odorant aglycones from odourless grape glycosidic aroma precursors. *Food Chem.* 2015;187:112–119.

199 aromatic compounds Piombino P, Genovese A, Esposito S, et al. Saliva from obese individuals suppresses the release of aroma compounds from wine. *PLoS ONE.* 2014;9(1):e85611. doi:10.1371/journal.pone.0085611.

199 periodontal disease Allaker RP. Investigations into the micro-ecology of oral malodour in man and companion animals. *J Breath Res.* 2010;4(1):017103.

200 many of these changes Stumpf RM, Wilson BA, Rivera A, et al. The primate vaginal microbiome: comparative context and implications for human health and disease. *Am J Phys Anthropol.* 2013;152(Suppl S57):119–134.

200 dysfunctional vaginal microbes Ravel J, Brotman RM, Gajer P, et al. Daily temporal dynamics of vaginal microbiota before, during and after episodes of bacterial vaginosis. *Microbiome.* 2013;1(1):29. doi:10.1186/2049-2618-1-29.

200 microbes within the vagina Nikolaitchouk N. *The Female Genital Tract Microbiota: Composition, Relation to Innate Immune Factors, and Effects of Contraceptives.* Vastra Frolunda, Sweden: University of Gothenburg; 2009. https://gupea.ub.gu.se/bitstream/2077/20102/1/gupea_2077_20102_1.pdf. Accessed August 2, 2015; Nelson DB, Bellamy S, Odibo A, et al. Vaginal symptoms and bacterial vaginosis (BV): how useful is self-report? Development of a screening tool for predicting BV status. *Epidemiol Infect.* 2007;135(8):1369–1375.

200 menstrual period Chaban B, Links MG, Jayaprakash TP, et al. Characterization of the vaginal microbiota of healthy Canadian women through the menstrual cycle. *Microbiome.* 2014;2:23. doi:10.1168/2049-2618-2-23.

200 changes in scent Srinivasan S, Fredricks DN. The human vaginal bacterial biota and bacterial vaginosis. *Interdiscip Perspect Infect Dis.* 2008;2008:750479. doi:10.1155.2008/750479.

200 phase of the menstrual cycle Cerda-Molina AL, Hernández-López L, de la O CE, et al. Changes in men's salivary testosterone and cortisol levels, and in sexual desire after smelling female axillary and vulvar scents. *Front Endocrinol (Lausanne).* 2013;4:159. doi:10.3389/fendo.2013.00159.

201 complexity and in smaller numbers Vence T. Parsing the penis microbiome. *The Scientist.* May 29, 2014. http://www.the-scientist.com/?articles.view/articleNo/40092/title/Parsing-the-Penis-Microbiome/. Accessed August 2, 2015.

201 their sexual activity Yandell K. Circumcision alters the penis microbiome. *The Scientist.* April 18, 2013. http://www.the-scientist.com/?articles.view/articleNo/35145/title/Circumcision-Alters-the-Penis-Microbiome/. Accessed August 2, 2015.

201 microbiome as well Vence T. Parsing the penis microbiome. *The Scientist.* May 29, 2014. http://www.the-scientist.com/?articles.view/articleNo/40092/title/Parsing-the-Penis-Microbiome/. Accessed August 2, 2015.

201 bacterial vaginosis in women Fishy smell around penis: causes and solutions. *Health Anxiety & Panic Headquarters Symptoms Explained by Doctors.* http://www.scarysymptoms.com/2012/01/fishy-smell-around-penis-causes-and.html. Accessed August 2, 2015.

201 human armpit odor Wongchoosuk C, Lutz M, Kerdcharoen T. Detection and classification of human body odor using an electronic nose. *Sensors (Basel).* 2009;9(9):7234–7249. doi:10.3390/S90907234.

201 exhaled air Bach JP, Gold M, Mengel D, et al. Measuring compounds in exhaled air to detect Alzheimer's disease and Parkinson's disease. *PLoS ONE.* 2015;10(7):e0132227. doi:10.1871/journal.pone.0132227.

201 detect cancer De Meij TG, Larbi IB, van der Schee MP, et al. Electronic nose can discriminate colorectal carcinoma and advanced adenomas by fecal volatile biomarker analysis: proof of principle study. *Int J Cancer.* 2014;134(5):1132–1138.

202 rather than later Arasaradnam RP, Covington JA, Harmston C, Nwokolo CU. Review article: next generation diagnostic modalities in gastroenterology—gas phase volatile compound biomarker detection. *Aliment Pharmacol Ther.* 2014;39(8):780–789.

202 just odor detection Pluznick JL, Protzko RJ, Gevorgyan H, et al. Olfactory receptor responding to gut microbiota-derived signals plays a role in renin secretion and blood pressure regulation. *Proc Natl Acad Sci USA.* 2013;110(11):4410–4415.

203 scents as anything else Svoboda E. Scents and sensibility. *Psychology Today.* January 1, 2008. https://www.psychologytoday.com/articles/200801/scents-and-sensibility. Accessed August 1, 2015.

203 immune response capabilities Svoboda E. Scents and sensibility. *Psychology Today.* January 1, 2008. https://www.psychologytoday.com/articles/200801/scents-and-sensibility. Accessed August 1, 2015.

203 MHC Bolnick DI, Snowberg LK, Caporaso JG, et al. Major histocompatibility complex class IIb polymorphism influences gut microbiota composition and diversity. *Mol Ecol.* 2014;23(19):4831–4845.

203 *Human Scent Evidence* Prada PA, Curran AM, Furton KG. *Human Scent Evidence.* Boca Raton, FL: CRC Press; 2014.

204 HLA type Lin P, Bach M, Asquith M, et al. HLA-B27 and human β2-microglobulin affect the gut microbiota of transgenic rats. *PLoS ONE.* 2014;9(8):e105684. doi:10.1371/journal.pone.0105684.

12. Superorganism Makeover

208 or *Ruminococcus* Arumugam M, Raes J, Pelletier E, et al. Enterotypes of the human gut microbiome. *Nature.* 2011;473(7346):174–180.

208 mucins and sugars Keim B. Gut-bacteria mapping finds three global varieties. *WIRED.* April 20, 2011. http://www.wired.com/2011/04/gut-bacteria-types/. Accessed Aug. 2, 2015; Bushman FD, Lewis JD, Wu GD. Diet, gut enterotypes and health: is there a link? *Nestle Nutr Inst Workshop Ser.* 2013;77:65–73.

208 **the other by *Prevotella*** Lim MY, Rho M, Song YM, et al. Stability of gut enterotypes in Korean monozygotic twins and their association with biomarkers and diet. *Sci Rep.* 2014;4:7348. doi:10.1038/srep07348.

208 **clusters of gut microbes** Knights D, Ward TL, McKinlay CE, et al. Rethinking "enterotypes." *Cell Host Microbe.* 2014;16(4):433–437.

209 **can improve health** Hill C, Guarner F, Reid G, et al. Expert consensus document. The International Scientific Association for Probiotics and Prebiotics consensus statement on the scope and appropriate use of the term probiotic. *Nat Rev Gastroenterol Hepatol.* 2014;11(8):506–514.

209 **gastrointestinal disease** Ritchie ML, Romanuk TN. A meta-analysis of probiotic efficacy for gastrointestinal diseases. *PLoS ONE.* 2012;7(4):e34938. doi:10.1071/journal.pone.0034938; Hill C, Guarner F, Reid G, et al. Expert consensus document. The International Scientific Association for Probiotics and Prebiotics consensus statement on the scope and appropriate use of the term probiotic. *Nat Rev Gastroenterol Hepatol.* 2014;11(8):506–514.

210 ***Gastroenterology & Hepatology*** Hill C, Guarner F, Reid G, et al. Expert consensus document. The International Scientific Association for Probiotics and Prebiotics consensus statement on the scope and appropriate use of the term probiotic. *Nat Rev Gastroenterol Hepatol.* 2014;11(8):506–514.

211 **from other animals** Chung H, Pamp SJ, Hill JA, et al. Gut immune maturation depends on colonization with a host-specific microbiota. *Cell.* 2012;149(7):1578–1593.

212 **your microbes** Kolho KL, Korpela K, Jaakkola T, et al. Fecal microbiota in pediatric inflammatory bowel disease and its relation to inflammation. *Am J Gastroenterol.* 2015;110(6):921–930; Yang T, Santisteban MM, Rodriguez V, et al. Gut dysbiosis is linked to hypertension. *Hypertension.* 2015;65(6):1331–1340; Huang YJ, Nariya S, Harris JM, et al. The airway microbiome in patients with severe asthma: associations with disease features and severity. *J Allergy Clin Immunol.* 2015;136(4):874–884.

213 **proton pump inhibitors** Marlicz W, Loniewski I, Grimes DS, Quigley EM. Nonsteroidal anti-inflammatory drugs, proton pump inhibitors, and gastrointestinal injury: contrasting interactions in the stomach and small intestine. *Mayo Clin Proc.* 2014;89(12):1699–1709.

213 **certain gut bacteria** Gut bacteria may affect whether a statin drug lowers cholesterol. Science News. *Science Daily.* October 14, 2011. http://www.science daily.com/releases/2011/10/111013184815.htm. Accessed August 3, 2015.

213 **adverse drug effects** Björklund M, Ouwehand AC, Forssten SD, et al. Gut microbiota of healthy elderly NSAID users is selectively modified with the administration of *Lactobacillus acidophilus* NCFM and lactitol. *Age (Dordr).* 2012;34(4):987–999.

214 **affected food preferences** Ursell LK, Knight R. Xenobiotics and the human gut microbiome: metatranscriptomics reveal the active players. *Cell Metab.* 2013;17(3):317–318. doi:10.1016/j.cmet.2013.02.013; McNulty NP, Yatsunenko T, Hsiao A, et al. The impact of a consortium of fermented milk strains on the gut microbiome of gnotobiotic mice and monozygotic twins. *Sci Transl Med.* 2011;3(106):106ra106.

214 **effectiveness considerably** German JB. The future of yogurt: scientific and regulatory needs. *Am J Clin Nutr.* 2014;99(5 Suppl):1271S–1278S.

214 reproductive tract Kumar N, Behera B, Sagiri SS, et al. Bacterial vaginosis: etiology and modalities of treatment—a brief note. *J Pharm Bioallied Sci.* 2011;3(4):496–503.

214 topical application Vaughn AR, Sivamani RK. Effects of fermented dairy products on skin: a systematic review. *J Altern Complement Med.* 2015;21(7):380–385.

215 inhibiting periodontal disease Terai T, Okumura T, Imai S, et al. Screening of probiotic candidates in human oral bacteria for the prevention of dental disease. *PLoS ONE.* 2015;10(6):e0128657. doi:10.1371/journal.pone.0128657.

215 prevention and therapeutics Zemanick ET, Sagel SD, Harris JK. The airway microbiome in cystic fibrosis and implications for treatment. *Curr Opin Pediatr.* 2011;23(3):319–324.

215 real internal struggle Festi D, Schiumerini R, Eusebi LH, et al. Gut microbiota and metabolic syndrome. *World J Gastroenterol.* 2014;20(43):16079–16094.

222 the Nurmi concept Nurmi E, Rantala M. New aspects of Salmonella infection in broiler production. *Nature.* 1973;241(5386):210–211.

223 specific prebiotic feeding Patterson JA, Burkholder KM. Application of prebiotics and probiotics in poultry production. *Poult Sci.* 2003;82(4):627–631.

223 good option Wolfenden AD, Pixley CM, Higgins JP, et al. Evaluation of spray application of a *Lactobacillus*-based probiotic on *Salmonella enteritidis* colonization in broiler chickens. *Int J Poult Sci.* 2007;6(7):493–496.

224 birds' microbiome Schneitz C. Competitive exclusion in poultry—30 years of research. *Food Control.* 2005;16(8):657–667.

224 under management stress Direct-fed microbials (probiotics) in calf diets. A BAMN Publication. Bovine Alliance on Management and Nutrition. http://www.aphis.usda.gov/animal_health/nahms/dairy/downloads/bamn/BAMN11_Probiotics.pdf. Accessed August 5, 2015.

225 from gaining a foothold Zhang WH, Jiang Y, Zhu QF, et al. Sodium butyrate maintains growth performance by regulating the immune response in broiler chickens. *Br Poult Sci.* 2011;52(3):292–301; Fang CL, Sun H, Wu J, et al. Effects of sodium butyrate on growth performance, haematological and immunological characteristics of weanling piglets. *J. Anim Physiol Anim Nutr.* 2014;98(4):680–685.

225 skewed as well Honneffer JB, Minamoto Y, Suchodolski JS. Microbiota alterations in acute and chronic gastrointestinal inflammation of cats and dogs. *World J Gastroenterol.* 2014;20(44):16489–16497; Suchodolski JS, Dowd SE, Wilke V, et al. 16S rRNA gene pyrosequencing reveals bacterial dysbiosis in the duodenum of dogs with idiopathic inflammatory bowel disease. *PLoS ONE.* 2012;7(6):e39333. doi:10.1371/journal.pone.0039333; Rodrigues Hoffmann A, Patterson AP, Diesel A, et al. The skin microbiome in healthy and allergic dogs. *PLoS ONE.* 2014;9(1):e83197. doi:10.1371/journal.pone.0083197.

225 diseases in dogs Strompfová V, Plachá I, Čobanová K, et al. Experimental addition of *Eleutherococcus senticosus* and probiotic to the canine diet. *Cent Eur J Biol.* 2012;7(3):436–447.

225 therapy, and standard care Grześkowiak Ł, Endo A, Beasley S, Salminen S. Microbiota and probiotics in canine and feline welfare. *Anaerobe.* 2015;34:14–23.

225 infections in pets Raditic DM. Complementary and integrative therapies for lower urinary tract diseases. *Vet Clin North Am Small Anim Pract.* 2015;45(4):857–878.

225 bowel disease in dogs Rossi G, Pengo G, Caldin M, et al. Comparison of microbiological, histological, and immunomodulatory parameters in response to treatment with either combination therapy with prednisone and metronidazole or probiotic VSL#3 strains in dogs with idiopathic inflammatory bowel disease. *PLoS ONE.* 2014;9(4):e94699. doi:10.1371/journal.pone.0094699.

225 inflammatory bowel disease Rossi G, Pengo G, Caldin M, et al. Comparison of microbiological, histological, and immunomodulatory parameters in response to treatment with either combination therapy with prednisone and metronidazole or probiotic VSL#3 strains in dogs with idiopathic inflammatory bowel disease. *PLoS ONE.* 2014;9(4):e94699. doi:10.1371/journal.pone .0094699.

226 healthy versus sick cats Suchodolski JS, Foster ML, Sohail MU, et al. The fecal microbiome in cats with diarrhea. *PLoS ONE.* 2015;10(5):e0127378. doi:10.1371/journal.pone.0127378.

226 renal diseases Cook AK. How to manage feline chronic diarrhea, part II: treatment. *DVM 360.* July 1, 2010. http://veterinarymedicine.dvm360.com/how -manage-feline-chronic-diarrhea-part-ii-treatment. Accessed August 4, 2015; Levy D. The use of probiotics in cats with renal failure. *The Nest.* http://pets .thenest.com/use-probiotics-cats-renal-failure-11104.html. Accessed August 4, 2015.

226 disease in cats Inflammatory bowel disease. Cornell University College of Veterinary Medicine. *The Feline Health Center.* http://www.vet.cornell.edu/ fhc/health_information/brochure_ibd.cfm. Accessed August 4, 2015.

227 *rhamnosus,* and *salivarius* Accepted claims about the nature of probiotic microorganisms in food. *Health Canada.* April 2009. http://www.hc-sc.gc.ca/ fn-an/label-etiquet/claims-reclam/probiotics_claims-allegations_probiotiques -eng.php. Accessed August 4, 2015.

13. To Be a Microbiome Whisperer

230 need to coexist Cani PD, Everard A. Talking microbes: when gut bacteria interact with diet and host organs. *Mol Nutr Food Res.* 2015. doi:10.1002/ mnfr.201500406.

231 with healthy controls Abrahamsson TR, Jakobsson HE, Andersson AF, et al. Low gut microbiota diversity in early infancy precedes asthma at school age. *Clin Exp Allergy.* 2014;44(6):842–850; Scher JU, Ubeda C, Artacho A, et al. Decreased bacterial diversity characterizes the altered gut microbiota in patients with psoriatic arthritis, resembling dysbiosis in inflammatory bowel disease. *Arthritis Rheumatol.* 2015;67(1):128–139; Shankar V, Hamilton MJ, Khoruts A, et al. Species and genus level resolution analysis of gut microbiota in *Clostridium difficile* patients following fecal microbiota transplantation. *Microbiome.* 2014;2(1):13. doi:10.1186/2049-2618-2-13; Hullar MA, Lampe JW. The gut microbiome and obesity. *Nestle Nutr Inst Workshop Ser.* 2012;73:67–79.

231 group of humans Clemente JC, Pehrsson EC, Blaser MJ, et al. The microbiome of uncontacted Amerindians. *Sci Adv.* 2015;1(3). pii: e1500183. doi:10.1126/ sciadv.1500183.

231 infectious diseases Hurtado AM, Lambourne CA, James P, et al. Human rights, biomedical science, and infectious diseases among South American

NOTES

indigenous groups. *Annu Rev Anthropol.* 2005;34(1):639–665; Hames R, Kuzara J. The nexus of Yanomamö growth, health and demography. In: Salzano FM, Magdalena Hurtado A, eds. *Lost Paradises and the Ethics of Research and Publication.* New York, NY: Oxford University Press; 2003:Chapter 7.

231 contact with outsiders Goering L. Invasion of gold miners threat to Brazilian tribe. *Chicago Tribune.* September 3, 1996. http://articles.chicagotribune.com/1996-09-03/news/9609030156_1_miners-yanomami-land-brazilian-tribe. Accessed September 14, 2015.

231 in recent decades Mancilha-Carvalho Jde J, Souza e Silva NA. The Yanomami Indians in the INTERSALT study. *Arq Bras Cardiol.* 2003;80(3):289–300; Manchila-Carvalho JJ, Crews DE. Lipid profiles of Yanomamo Indians of Brazil. *Prev Med.* 1990;19(1):66–75.

231 westernized lifestyle Hidalgo G, Marini E, Sanchez W, et al. The nutrition transition in the Venezuelan Amazonia: increased overweight and obesity with transculturation. *Am J Human Biol.* 2014;26(5):710–712.

232 gastrointestinal cancers Kahouli I, Tomaro-Duchesneau C, Prakash S. Probiotics in colorectal cancer (CRC) with emphasis on mechanisms of action and current perspectives. *J Med Microbiol.* 2013; 62(Pt 8):1107–1123.

232 patient prognosis Mima K, Nishihara R, Qian ZR, et al. *Fusobacterium nucleatum* in colorectal carcinoma tissue and patient prognosis. *Gut.* 2015. pii: gutjnl-2015-310101.

232 gut microbes overall Gao Z, Guo B, Gao R, et al. Probiotics modify human intestinal mucosa-associated microbiota in patients with colorectal cancer. *Mol Med Rep.* 2015;12(4):6119–6127.

233 countries in Asia Nakayama J, Watanabe K, Jiang J, et al. Diversity in gut bacterial community of school-age children in Asia. *Sci Rep.* 2015;5:8397.

233 counterparts in Bangkok Ruengsomwong S, Korenori Y, Sakamoto N, et al. Senior Thai fecal microbiota comparison between vegetarians and non-vegetarians using PCR-DGGE and real-time PCR. *J Microbiol Biotechnol.* 2014;24(8):1026–1033.

234 in the northeast La-Ongkham O, Nakphaichit M, Leelavatcharamas V, et al. Distinct gut microbiota of healthy children from two different geographic regions of Thailand. *Arch Microbiol.* 2015;197(4):561–573.

234 features of the microbiome Kemppainen KM, Ardissone AN, Davis-Richardson AG, et al. Early childhood gut microbiomes show strong geographic differences among subjects at high risk for type 1 diabetes. *Diabetes Care.* 2015;38(2):329–332.

234 Germany, and Africa Li J, Quinque D, Horz HP, et al. Comparative analysis of the human saliva microbiome from different climate zones: Alaska, Germany, and Africa. *BMC Microbiol.* 2014;14:316. doi:10.1186/S12866-014-0316-1.

236 increases significantly Rutherford ST, Bassler BL. Bacterial quorum sensing: its role in virulence and possibilities for its control. *Cold Spring Harb Perspect Med.* 2012;2(11). pii: a012427. doi:10.1101/cshperspect.a012427.

236 study from the Netherlands Sturme MH, Francke C, Siezen RJ, et al. Making sense of quorum sensing in lactobacilli: a special focus on *Lactobacillus plantarum* WCFS1. *Microbiology.* 2007;153(Pt 12):3939–3947.

236 At least one study Viswanathan VK. Sensing bacteria, without bitterness? *Gut Microbes.* 2013;4(2):91–93.

319

237 microbiome management LaSarre B, Federle MJ. Exploiting quorum sensing to confuse bacterial pathogens. *Microbiol Mol Biol Rev.* 2013;77(1):73–111.

237 enzyme called acylase Ivanova K, Fernandes MM, Mendoza E, Tzanov T. Enzyme multilayer coatings inhibit *Pseudomonas aeruginosa* biofilm formation on urinary catheters. *Appl Microbiol Biotechnol.* 2015;99(10):4373–4385.

237 microbes within us Aggarwal C, Jimenez JC, Lee H, et al. Identification of quorum-sensing inhibitors disrupting signaling between Rgg and short hydrophobic peptides in Streptococci. *MBio.* 2015;6(3):e00393–15. doi:10.1128/mBio .00393-15.

238 an intriguing study Hsiao A, Ahmed AM, Subramanian S, et al. Members of the human gut microbiota involved in recovery from *Vibrio cholerae* infection. *Nature.* 2014;515(7527):423–426.

239 subvert their functions Barrangou R, Horvath P. The CRISPR system protects microbes against phages, plasmids. *Microbe.* 2009;4(5):224–230.

239 Centre for Infection Research Doudna JA, Charpentier E. Genome editing. The new frontier of genome engineering with CRISPR-Cas9. *Science.* 2014;346(6213):1258096.

239 Carl Zimmer in *Quanta Magazine* Zimmer C. Breakthrough DNA editor borne of bacteria. *Quanta Magazine.* February 6, 2015. https://www.quanta magazine.org/20150206-crispr-dna-editor-bacteria/. Accessed October 11, 2015.

240 capacity of memory Virus-cutting enzyme helps bacteria remember a threat. *Science News.* February 20, 2015. http://newswire.rockefeller.edu/2015 /02/20/virus-cutting-enzyme-helps-bacteria-remember-a-threat/. Accessed August 6, 2015.

240 intact virus attack Hynes AP, Villion M, Moineau S. Adaptation in bacterial CRISPR-Cas immunity can be driven by defective phages. *Nat Commun.* 2014;5:4399. doi:10.1038/ncomms5399.

240 host bacterial genome Paez-Espino D, Morovic W, Sun CL, et al. Strong bias in the bacterial CRISPR elements that confer immunity to phage. *Nat Commun.* 2013;4:1430. doi:10.1038/ncomms2440.

240 area of the chromosome Levy A, Goren MG, Yosef I, et al. CRISPR adaptation biases explain preference for acquisition of foreign DNA. *Nature.* 2015;520(7548):505–510.

241 cell envelope physiology Ratner HK, Sampson TR, Weiss DS. I can see CRISPR now, even when phage are gone: a view on alternative CRISPR-Cas functions from the prokaryotic envelope. *Curr Opin Infect Dis.* 2015;28(3):267–274.

241 better battle disease The CRISPR revolution. *Science AAAS.* http://www .sciencemag.org/site/extra/crispr/?intcmp=HP-COLLECTION-PROMO-crispr. Accessed August 7, 2015.

14. Your Brain on Microbes

242 running the world Jensen T. Democrats and Republicans differ on conspiracy theory beliefs. *Public Policy Polling.* April 2, 2013. http://www.publicpolicy polling.com/pdf/2011/PPP_Release_National_ConspiracyTheories_040213.pdf. Accessed August 11, 2015.

243 short-chain fatty acids Dinan TG, Stilling RM, Stanton C, Cryan JF. Collective unconscious: how gut microbes shape human behavior. *J Psychiatr Res.* 2015;63:1–9.

243 parasites and pathogens Stilling RM, Dinan TG, Cryan JF. The brain's Geppetto-microbes as puppeteers of neural function and behaviour? *J Neurovirol.* 2015. doi:10.1007/s13365-015-0355-x.

243 serotonin levels Jackson AC. Diabolical effects of rabies encephalitis. *J Neurovirol.* 2015. doi:10.1007/s13365-015-0351-1.

243 microbe's intent Stilling RM, Dinan TG, Cryan JF. The brain's Geppetto-microbes as puppeteers of neural function and behaviour? *J Neurovirol.* 2015. doi: 10.1007/s13365-015-0355-x.

243 "second brain" Hadhazy A. Think twice: how the gut's "second brain" influences mood and well-being. *Scientific American.* February 12, 2010. http://www.scientificamerican.com/article/gut-second-brain/. Accessed August 11, 2015.

244 as described in Part One Stilling RM, Bordenstein SR, Dinan TG, Cryan JF. Friends with social benefits: host-microbe interactions as a driver of brain evolution and development? *Front Cell Infect Microbiol.* 2014;4:147. http://dx.doi.org/10.3389/fcimbi2014.00147.

244 do precisely that Alcock J, Maley CC, Aktipis CA. Is eating behavior manipulated by the gastrointestinal microbiota? Evolutionary pressures and potential mechanisms. *BioEssays.* 2014;36(10):940–949.

244 preventing NCDs Manach C, Scalbert A, Morand C, et al. Polyphenols: food sources and bioavailability. *Am J Clin Nutr.* 2004;79(5):727–747.

244 these chemicals, too Cardona F, Andrés-Lacueva C, Tulipan, S, et al. Benefits of polyphenols on gut microbiota and implications in human health. *J Nutr Biochem.* 2013;24(8):1415–1422.

244 metabolites in their urine Alcock J. Maley CC, Aktipis CA. Is eating behavior manipulated by the gastrointestinal microbiota? Evolutionary pressures and potential mechanisms. *BioEssays.* 2014;36(10): 940–949; Rezzi S, Ramadan Z, Martin FP, et al. Human metabolic phenotypes link directly to specific dietary preferences in healthy individuals. *J Proteome Res.* 2007;6(11):4469–4477.

245 biomarker for obesity De Filippo C, Cavalieri D, Di Paola M, et al. Impact of diet in shaping gut microbiota revealed by a comparative study in children from Europe and rural Africa. *Proc Natl Acad Sci USA.* 2010;107(33):14691–14696.

245 higher B-to-F ratio Norris V, Molina F, Gewirtz AT. Hypothesis: bacteria control host appetites. *J Bacteriol.* 2013;195(3):411–416.

245 crave dietary fiber Alcock J, Maley CC, Aktipis CA. Is eating behavior manipulated by the gastrointestinal microbiota? Evolutionary pressures and potential mechanisms. *BioEssays.* 2014;36(10):940–949.

245 precisely what they want Norris V, Molina F, Gewirtz AT. Hypothesis: bacteria control host appetites. *J Bacteriol.* 2013;195(3):411–416.

246 from the gut to the liver Lyte M. Microbial endocrinology: host-microbiota neuroendocrine interactions influencing brain and behavior. *Gut Microbes.* 2014;5(3):381–389.

246 and additional bacteria Lyte M. Microbial endocrinology: host-microbiota neuroendocrine interactions influencing brain and behavior. *Gut Microbes.* 2014;5(3):381–389; Dinan TG, Stilling RM, Stanton C, Cryan JF. Collective unconscious: how gut microbes shape human behavior. *J Psychiatr Res.* 2015;63:1–

9; Thomas CM, Hong T, van Pijkeren JP, et al. Histamine derived from probiotic *Lactobacillus reuteri* suppresses TNF via modulation of PKA and ERK signaling. *PLoS ONE.* 2012;7(2):e31951. doi:10.1371/journal.pone.0031951.

246 bladder control Hadhazy A. Think twice: how the gut's "second brain" influences mood and well-being. *Scientific American.* February 12, 2010. http://www.scientificamerican.com/article/gut-second-brain/. Accessed August 11, 2015; Berger M, Gray JA, Roth BL. The expanded biology of serotonin. *Annu Rev Med.* 2009;60:355–366.

246 gut bacteria produce Yano JM, Yu K, Donaldson GP, et al. Indigenous bacteria from the gut microbiota regulate host serotonin biosynthesis. *Cell.* 2015;161(2):264–276; Ridaura V, Belkaid Y. Gut microbiota: the link to your second brain. *Cell.* 2015;161(2):193–194.

247 behavioral characteristics Stilling RM, Dinan TG, Cryan JF. Microbial genes, brain & behaviour—epigenetic regulation of the gut-brain axis. *Genes Brain Behav.* 2014;13(1):69–86; Borre YE, Moloney RD, Clarke G, et al. The impact of microbiota on brain and behavior: mechanisms & therapeutic potential. *Adv Exp Med Biol.* 2014;817:373–403.

247 connected to neurobehavior Stilling RM, Ryan FJ, Hoban AE, et al. Microbes & neurodevelopment—absence of microbiota during early life increases activity-related transcriptional pathways in the amygdala. *Brain Behav Immun.* 2015;50:209–220.

247 social cognition Desbonnet L, Clarke G, Shanahan F, et al. Microbiota is essential for social development in the mouse. *Mol Psychiatry.* 2014;19(2):146–148.

247 microbes between individuals Stilling RM, Bordenstein SR, Dinan TG, Cryan JF. Friends with social benefits: host-microbe interactions as a driver of brain evolution and development? *Front Cell Infect Microbiol.* 2014 ;4:147.

247 include our microbes Dinan TG, Stilling RM, Stanton C, Cryan, JF. Collective unconscious: how gut microbes shape human behavior. *J Psychiatr Res.* 2015;63:1–9.

247 cognitive deficits Desbonnet L, Clarke G, Traplin A, et al. Gut microbiota depletion from early adolescence in mice: implications for brain and behaviour. *Brain Behav Immun.* 2015;48:165–173.

247 autistic children Dinan TG, Stilling RM, Stanton C, Cryan JF. Collective unconscious: how gut microbes shape human behavior. *J Psychiatr Res.* 2015;63:1–9.

248 central-nervous-system inflammation Galland L. The gut microbiome and the brain. *J Med Food.* 2014;17(12):1261–1272.

248 reduce systemic inflammation Groeger D, O'Mahony L, Murphy EF, et al. *Bifidobacterium infantis* 35624 modulates host inflammatory processes beyond the gut. *Gut Microbes.* 2013;4(4):325–339.

248 June 2014 Brooks M. Top 10 most prescribed, top-selling drugs. Medscape Medical News, *WebMD.* August 5, 2014. http://www.webmd.com/news/20140805/top-10-drugs. Accessed August 13, 2015.

248 suicide in the young Aripiprazole. *MedlinePlus.* http://www.nlm.nih.gov/medlineplus/druginfo/meds/a603012.html. Accessed August 17, 2015.

249 rose by double digits Pritchard C, Rosenorn-Lanng E. Neurological deaths of American adults (55–74) and the over 75's by sex compared with 20 Western countries 1989–2010: cause for concern. *Surg Neurol Int.* 2015;6:123.

249 another dementia 2015 Alzheimer's disease facts and figures. *Alzheimer's Assocation.* http://www.alz.org/facts/. Accessed August 14, 2015.

249 **autism spectrum disorder** *Centers for Disease Control and Prevention.* March 27, 2014. http://www.gov/nchs/data/nhsr/nhsr087.pdf. Accessed November 18, 2015.

249 **$247 billion each year** Children's mental health—new report. *Centers for Disease Control and Prevention.* http://www.cdc.gov/features/childrensmental health/. Accessed August 14, 2015.

250 **differ between strains** Savignac HM, Tramullas M, Kiely B, et al. Bifidobacteria modulate cognitive processes in an anxious mouse strain. *Behav Brain Res.* 2015;287:59–72.

250 **antianxiety action in mice** Savignac HM, Kiely B, Dinan TG, Cryan JF. Bifidobacteria exert strain-specific effects on stress-related behavior and physiology in BALB/c mice. *Neurogastroenterol Motil.* 2014;26(11):1615–1627.

250 **stressed students** Langkamp-Henken B, Rowe CC, Ford AL, et al. *Bifidobacterium bifidum* R0071 results in a greater proportion of healthy days and a lower percentage of academically stressed students reporting a day of cold/flu: a randomised, double-blind, placebo-controlled study. *Br J Nutr.* 2015;113(3):426–434. doi:10.1017/S0007114514003997.

250 **reduced social anxiety** Hilimire MR, DeVylder JE, Forestell CA. Fermented foods, neuroticism, and social anxiety: an interaction model. *Psychiatry Res.* 2015;228(2):203–208.

250 **compared with controls** Mohammadi AA, Jazayeri S, Khosravi-Darani K, et al. The effects of probiotics on mental health and hypothalamic-pituitary-adrenal axis: a randomized, double-blind, placebo-controlled trial in petrochemical workers. *Nutr Neurosci.* April 16, 2015. doi:10.1179/1476830515Y .0000000023.

251 **needed metabolites** Tarr AJ, Galley JD, Fisher SE, et al. The prebiotics 3'Sialyllactose and 6'Sialyllactose diminish stressor-induced anxiety-like behavior and colonic microbiota alterations: evidence for effects on the gut-brain axis. *Brain Behav Immun.* 2015;50:166–177.

251 ***Bifidobacterium longum*** Yu ZT, Chen C, Newburg DS. Utilization of major fucosylated and sialylated human milk oligosaccharides by isolated human gut microbes. *Glycobiology.* 2013;23(11):1281–1292.

251 **poststress behavioral testing** Tarr AJ, Galley JD, Fisher SE, et al. The prebiotics 3'Sialyllactose and 6'Sialyllactose diminish stressor-induced anxiety-like behavior and colonic microbiota alterations: evidence for effects on the gut-brain axis. *Brain Behav Immun.* 2015;50:166–177.

251 **attention and focus** Schmidt K, Cowen PJ, Harmer CJ, et al. Prebiotic intake reduces the waking cortisol response and alters emotional bias in healthy volunteers. *Psychopharmacology (Berl).* 2015;232(10):1793–1801.

251 **supplemented formulas** Vandenplas Y, Zakharova I, Dmitrieva Y. Oligosaccharides in infant formula: more evidence to validate the role of prebiotics. *Br J Nutr.* 2015;113(9):1339–1344.

251 **prevalence of colic** Giovannini M, Verduci E, Gregori D, et al. Prebiotic effect of an infant formula supplemented with galacto-oligosaccharides: randomized multicenter trial. *J Am Coll Nutr.* 2014;33(5):385–393.

252 **the hippocampus** Dinan TG, Stilling RM, Stanton C, Cryan JF. Collective unconscious: how gut microbes shape human behavior. *J. Psychiatr Res.* 2015;63:1–9.

15. Will You Do No Harm?

254 risk of NCDs Bisphenol A (BPA). *National Institute of Environmental Health Sciences.* https://www.niehs.nih.gov/health/topics/agents/sya-bpa/. Accessed August 8, 2015; Toxicological profile for Bisphenol A. *California Environmental Protection Agency.* September 2009. http://www.opc.ca.gov/webmaster/ftp/project_pages/MarineDebris_OEHHA_ToxProfiles/Bisphenol%20A%20Final.pdf. Accessed August 8, 2015.

254 to our immune system Braniste V, Jouault A, Gaultier E, et al. Impact of oral bisphenol A at reference doses on intestinal barrier function and sex differences after perinatal exposure in rats. *Proc Natl Acad Sci USA.* 2010;107(1):448–453.

254 life stage, infancy Bisphenol A: EU ban on baby bottles to enter into force tomorrow. European Commission. January 2011. http://europa.eu/rapid/press-release_IP-11-664_en.htm. Accessed August 8, 2015.

254 much consumer protest Summary of bisphenol A (BPA) regulation (2nd edition). *Modern Testing Services.* May 29, 2013. http://www.mts-global.com/en/technical_update/CPIE-018-13.html. Accessed August 8, 2015; Walsh B. Why the FDA hasn't banned potentially toxic BPA (yet). *Time.* April 3, 2012. http://content.time.com/time/health/article/0,8599,2110902,00.html. Accessed August 8, 2015.

254 and who lags Environment. *European Union.* http://europa.eu/pol/env/index_en.htm. Accessed August 8, 2015.

254 inflammation-driven obesity Food additives alter gut microbes, cause diseases in mice. *NIH Research Matters.* March 16, 2015. http://www.nih.gov/news-events/nih-research-matters/food-additives-alter-gut-microbes-cause-diseases-mice. Accessed August 8, 2015.

254 biological plausibility are clear Food additives alter gut microbes, cause diseases in mice. *NIH Research Matters.* March 16, 2015. http://www.nih.gov/news-events/nih-research-matters/food-additives-alter-gut-microbes-cause-diseases-mice. Accessed August 8, 2015.

255 interactions are extensive Wilson ID, Nicholson JK. The modulation of drug efficacy and toxicity by the gut microbiome. In: Kochhar S, Martin F-P, eds. *Metabonomics and Gut Microbiota in Nutrition and Disease.* New York, NY: Humana Press; 2015.

256 true in the US Myhr AI. The precautionary principle in GMO regulations. In: Traavic T, Ching LL, eds. *Biosafety First.* Trondheim, Norway: Tapir Academic; 2007:Chapter 29; Germany joins ranks of anti-GMO countries. *EurActiv.com.* April 15, 2009. http://www.euractiv.com/cap/germany-joins-ranks-anti-gmo-cou-news-221725. Accessed August 17, 2015.

256 types of biofilms Lima IS, Baumeier NC, Rosa RT, et al. Influence of glyphosate in planktonic and biofilm growth of *Pseudomonas aeruginosa. Braz J Microbiol.* 2014;45(3):971–975.

256 support foraging animals Druille M, Cabello MN, Garcia Parisi PA, et al. Glyphosate vulnerability explains changes in root-symbionts propagules viability in pampean grasslands. *Agric Ecosys Environ* 2015;202:48–55. doi:10.1016/j.agee.2014.12.017.

256 pathogenic to humans Kurenbachk B, Marjoshi D, Amábile-Cuevas CF, et al. Sublethal exposure to commercial formulations of the herbicides dicamba, 2,4-dichlorophenoxyacetic acid, and glyphosate cause changes in antibiotic susceptibility in *Escherichia coli* and *Salmonella enterica* serovar typhimurium. *mBio.* 2015;6(2):e00009–15. doi:10.1128mBio.00009-15.

256 **helpful commensal bacteria** Shehata AA, Schrödl W, Aldin AA, et al. The effect of glyphosate on potential pathogens and beneficial members of poultry microbiota in vitro. *Curr Microbiol.* 2013;66(4):350–358.

257 ***Clostridium botulinum*** Schrödl W, Krüger S, Konstantinova-Müller T, et al. Possible effects of glyphosate on Mucorales abundance in the rumen of dairy cows in Germany. *Curr Microbiol.* 2014;69(6):817–823; Ackermann W, Coenen M, Schrödl W, et al. The influence of glyphosate on the microbiota and production of botulinum neurotoxin during ruminal fermentation. *Curr Microbiol.* 2015;70(3):374–382; Krüger M, Shehata AA, Schrödl W, Rodloff A. Glyphosate suppresses the antagonistic effect of *Enterococcus* spp. on *Clostridium botulinum. Anaerobe.* 2013;20:74–78.

259 **the FDA handles them** Venugopalan V, Shriner KA, Wong-Beringer A. Regulatory oversight and safety of probiotic use. *Emerg Infect Dis.* 2010:16(11). http://wwwnc.cdc.gov/eid/article/16/11/10-0574_article. Accessed August 10, 2015.

259 **any medical use** Lactobacillus acidophilus. *University of Maryland Medical Center.* https://umm.edu/health/medical/altmed/supplement/lactobacillus-acidophilus. Accessed August 10, 2015.

259 **company shut down** FDA: DDS probiotic products seized. *FDA News Release. FDA.* June 7, 2011. http://www.fda.gov/NewsEvents/Newsroom/PressAnnounce ments/ucm258155.htm. Accessed August 10, 2015.

259 **good healthy food** Bubnov RV, Spivak MY, Lazarenko LM, et al. Probiotics and immunity: provisional role for personalized diets and disease prevention. *EPMA J.* 2015;6(1):14.

259 **offer one option** Mueller NT, Bakacs E, Combellick J, et al. The infant microbiome development: mom matters. *Trends Mol Med.* 2015;21(2):109–117.

260 **shift your microbiome** Fasano A. Intestinal permeability and its regulation by zonulin: diagnostic and therapeutic implications. *Clin Gastroenterol Hepatol.* 2012;10(10):1096–1100; Venkatesh M, Mukherjee S, Wang H, et al. Symbiotic bacterial metabolites regulate gastrointestinal barrier function via the xenobiotic sensor PXR and Toll-like receptor 4. *Immunity.* 2014;41(2):296–310.

261 **predominant group** Mathur R, Barlow GM. Obesity and the microbiome. *Expert Rev Gastroenterol Hepatol.* 2015;9(8):1087–1099.

261 **periodontal disease** Flichy-Fernández AJ, Ata-Ali J, Alegre-Domingo T, et al. The effect of orally administered probiotic *Lactobacillus reuteri*-containing tablets in peri-implant mucositis: a double-blind randomized controlled trial. *J Periodontal Res.* 2015;50(6):775–785.

262 **bacterial vaginosis** Martínez-Peña MD, Castro-Escarpulli G, Aguilera-Arreola MG. *Lactobacillus* species isolated from vaginal secretions of healthy and bacterial vaginosis-intermediate Mexican women: a prospective study. *BMC Infect Dis.* 2013;13:189. doi:10.1186/1471-2334-13-189.

262 **respiratory tract priming** Percopo CM, Rice TA, Brenner TA, et al. Immunobiotic *Lactobacillus* administered post-exposure averts the lethal sequelae of respiratory virus infection. *Antiviral Res.* 2015;121:109–119.

262 **prevent damaging inflammation** Zelaya H, Tada A, Vizoso-Pinto MG, et al. Nasal priming with immunobiotic *Lactobacillus rhamnosus* modulates inflammation-coagulation interactions and reduces influenza virus-associated pulmonary damage. *Inflamm Res.* 2015;64(8):589–602.

262 ultraviolet radiation Satoh T, Murata M, Iwabuchi N, et al. Effect of *Bifidobacterium breve* B-3 on skin photoaging induced by chronic UV irradiation in mice. *Benef Microbes.* 2015;6(4):497–504.

262 after UV exposure Ra J, Lee DE, Kim SH, et al. Effect of oral administration of *Lactobacillus plantarum* HY7714 on epidermal hydration in ultraviolet B-irradiated hairless mice. *J Microbiol Biotechnol.* 2014;24(12):1736–1743.

262 from drying out Nodake Y, Matsumoto S, Miura R, et al. Pilot study on novel skin care method by augmentation with *Staphylococcus epidermidis*, an autologous skin microbe—a blinded randomized clinical trial. *J Dermatol Sci.* 2015;79(2):119–126.

265 *Leuconostoc mesenteroides* Beganović J, Pavunc AL, Gjuračić K, et al. Improved sauerkraut production with probiotic strain *Lactobacillus plantarum* L4 and *Leuconostoc mesenteroides* LMG 7954. *J Food Sci.* 2011;76(2):M124–129.

267 bacteria as well Jung JY, Lee SH, Kim JM, et al. Metagenomic analysis of kimchi, a traditional Korean fermented food. *Appl Environ Microbiol.* 2011;77(7):2264–2274.

267 attributed to its consumption Vīna I, Semjonovs P, Linde R, Deniņa I. Current evidence on physiological activity and expected health effects of kombucha fermented beverage. *J Med Food.* 2014;17(2):179–188. doi:10.1089/jmf .2013.0031.

268 degradation of histamine Tsai Y-H. Degradation of histamine by *Lactobacillus plantarum* D103 isolated from miso, a fermented soybean food. International Association for Food Protection. *2015 Annual Conference.* Abstract #P2-215.

268 antioxidant properties Watanabe N, Fujimoto K, Aoki H. Antioxidant activities of the water-soluble fraction in tempeh-like fermented soybean (GABA-tempeh). *Int J Food Sci Nutr.* 2007;58(8):577–587.

268 *Saccharomyces cerevisiae* Dlusskaya E, Jänsch A, Schwab C, Gänzle M. Microbial and chemical analysis of a kvass fermentation. *Eur Food Res Technol.* 2008;227(1):261–266.

269 also prevalent Osvik RD, Sperstad S, Breines E, et al. Bacterial diversity of aMasi, a South African fermented milk product, determined by clone library and denaturing gradient gel electrophoresis analysis. *Afr J Microbiol Res.* 2013;7(32):4146–4158. doi:10.5897/AJMR12.2317.

269 *Cronobacter, Klebsiella* Puerari C, Magalhães-Guedes KT, Schwan RF. Physicochemical and microbiological characterization of chicha, a rice-based fermented beverage produced by Umutina Brazilian Amerindians. *Food Microbiol.* 2015;46:210–217.

270 severity of colds Hughes C, Davoodi-Semiromi Y, Colee JC, et al. Galactooligosaccharide supplementation reduces stress-induced gastrointestinal dysfunction and days of cold or flu: a randomized, double-blind, controlled trial in healthy university students. *Amer J Clin Nutr.* 2011;93(6):1305–1311.

270 among the elderly Get active. Let's move! letsmove.gov. http://www.letsmove .gov/get-active. Accessed August 9, 2015; Physical activity. Workplace Health Promotion. Centers for Disease Control and Prevention. http://www.cdc.gov/ workplacehealthpromotion/implementation/topics/physical-activity.html. Accessed August 9, 2015; Belza B, PRC-HAN Physical Activity Conference Planning Workgroup. Moving ahead: strategies and tools to plan, conduct, and maintain effective community-based physical activity programs for older adults: a brief guide. *Centers for Disease Control and Prevention*. Atlanta, GA; 2007. http://www.cdc.gov/aging/pdf/community-based_physical_activity _programs_for_older_adults.pdf. Accessed August 9, 2015.

270 **useful gut microbes** Bermon S, Petriz B, Kajėnienė A, et al. The microbiota: an exercise immunology perspective. *Exerc Immunol Rev.* 2015;21:70–79.

270 **independent of diet** Kang SS, Jeraldo PR, Kurti A, et al. Diet and exercise orthogonally alter the gut microbiome and reveal independent associations with anxiety and cognition. *Mol Neurodegener.* 2014;9:36. doi:10.1186/1750-1326 -9-36.

270 **high-fat diet** Evans CC, LePard K, Kwak JW, et al. Exercise prevents weight gain and alters the gut microbiota in a mouse model of high fat diet-induced obesity. *PLoS ONE.* 2014;9(3):e92193. doi:10.1371/journal.pone.0092193.

270 **more anti-inflammatory** Allen JM, Berg Miller ME, Pence BD, et al. Voluntary and forced exercise differentially alters the gut microbiome in C57BL/6J mice. *J Appl Physiol.* 2015;118(8):1059–1066.

271 **gut microbiome** Choi JJ, Eum SY, Rampersaud E, et al. Exercise attenuates PCB-induced changes in the mouse gut microbiome. *Environ Health Perspect.* 2013;121(6):725–730.

272 **"what good is it"** Maurois A. Hopkins G, trans. *The Life of Sir Alexander Fleming, Discoverer of Penicillin.* New York, NY: E.P. Dutton & Co.; 1959; Brown, K. *Penicillin Man: Alexander Fleming and the Antibiotic Revolution.* The History Press, Kindle version; 2013.

Resources for Probiotics

273 **the latest developments** FDA reconsiders IND requirement for food research. International Scientific Association for Probiotics and Prebiotics. *ISAPP Annual Meeting.* http://www.isapp.net/Home. Accessed August 10, 2015.

273 **prebiotics on their website** FDA reconsiders IND requirement for food research. International Scientific Association for Probiotics and Prebiotics. *ISAPP Annual Meeting.* http://www.isapp.net/Probiotics-and-Prebiotics/Re sources. Accessed August 10, 2015.

273 **educational organization** Probiotics for GI Health in 2012: issues and updates. *Primary Issues.* 2012. http://www.isapp.net/Portals/0/docs/News/ merenstein%20sanders%20CME%20Probiotics.pdf. Accessed August 10, 2015.

273 **available in Canada** Skokovic-Sunjic D. Clinical guide to probiotic supplements available in Canada (2015 edition). Indications, dosage forms, and clinical evidence to date. *Alliance for Education on Probiotics.* http://www.isapp .net/Portals/0/docs/clincial%20guide%20canada.pdf. Accessed August 10, 2015.

273 **studies and reviews** Varankovich NV, Nickerson MT, Korber DR. Probiotic-based strategies for therapeutic and prophylactic use against multiple gastrointestinal diseases. *Front Microbiol.* 2015;6:685. doi:10.3389/fmicb.2015.00685; Bubnov RV, Spivak MY, Lazarenko LM, et al. Probiotics and immunity: provisional role for personalized diets and disease prevention. *EPMA J.* 2015;6(1):14.

274 ***Akkermansia muciniphila*** Dao MC, Everard A, Aron-Wisnewsky J, et al. *Akkermansia muciniphila* and improved metabolic health during a dietary intervention in obesity: relationship with gut microbiome richness and ecology. *Gut.* June 22, 2015. pii: gutjnl-2014-308778.

274 **Bifidobacterium animalis** Centanni M, Turroni S, Rampelli S, et al. *Bifidobacterium animalis* ssp. lactis BIo7 modulates the tumor necrosis factor alpha-dependent imbalances of the enterocyte-associated intestinal microbiota fraction. *FEMS Microbiol Lett.* 2014;357(2):157–163.

274 **Bifidobacterium breve** Izumi H, Minegishi M, Sato Y, et al. *Bifidobacterium breve* alters immune function and ameliorates DSS-induced inflammation in weanling rats. *Pediatr Res.* 2015;78(4):407–416.

274 **Bifidobacterium infantis** Zuo L, Yuan KT, Yu L, et al. *Bifidobacterium infantis* attenuates colitis by regulating T cell subset responses. *World J Gastroenterol.* 2014;20(48):18316–18329.

274 **Bifidobacterium longum** Elian SD, Souza EL, Vieira AT, et al. *Bifidobacterium longum* subsp. infantis BB-02 attenuates acute murine experimental model of inflammatory bowel disease. *Benef Microbes.* 2015;6(3):277–286.

274 **Enterococcus durans** Avram-Hananel L, Stock J, Parlesak A, et al. *E durans* strain M4-5 isolated from human colonic flora attenuates intestinal inflammation. *Dis Colon Rectum.* 2010;53(12):1676–1686.

274 **Enterococcus faecalis** Wang S, Hibberd ML, Pettersson S, Lee YK. *Enterococcus faecalis* from healthy infants modulates inflammation through MAPK signaling pathways. *PLoS ONE.* 2014;9(5):e97523. doi:10.1371/journal.pone.0097523.

274 **Faecalibacterium prausnitzii** Quévrain E, Maubert MA, Michon C, et al. Identification of an anti-inflammatory protein from *Faecalibacterium prausnitzii*, a commensal bacterium deficient in Crohn's disease. *Gut.* June 4, 2015. pii: gutjnl-2014-307649.

274 **Lactobacillus acidophilus** Inoue Y, Kambara T, Murata N, et al. Effects of oral administration of *Lactobacillus acidophilus* L-92 on the symptoms and serum cytokines of atopic dermatitis in Japanese adults: a double-blind, randomized, clinical trial. *Int Arch Allergy Immunol.* 2014;165(4):247–254.

274 **Lactobacillus amylovorus** Finamore A, Roselli M, Imbinto A, et al. *Lactobacillus amylovorus* inhibits the TLR4 inflammatory signaling triggered by enterotoxigenic *Escherichia coli* via modulation of the negative regulators and involvement of TLR2 in intestinal Caco-2 cells and pig explants. *PLoS ONE.* 2014;9(4):e94891. doi:10.1371/journal.pone.0094891.

274 **Lactobacillus casei** Alipour B, Homayouni-Rad A, Vaghef-Mehrabany E, et al. Effects of *Lactobacillus casei* supplementation on disease activity and inflammatory cytokines in rheumatoid arthritis patients: a randomized double-blind clinical trial. *Int J Rheum Dis.* 2014;17(5):519–527.

274 **Lactobacillus fermentum** Jin P, Chen Y, Lv L, et al. *Lactobacillus fermentum* ZYLo401 attenuates lipopolysaccharide-induced Hepatic TNF-α expression and liver injury via an IL-10- and PGE2-EP4-dependent mechanism. *PLoS ONE.* 2015;10(5):e0126520. doi:10.1371/journal.pone.0126520.

274 **Lactobacillus gasseri** Spaiser SJ, Culpepper T, Nieves C Jr, et al. *Lactobacillus gasseri* KS-13, *Bifidobacterium bifidum* G9-1, and *Bifidobacterium longum* MM-2 ingestion induces a less inflammatory cytokine profile and a potentially beneficial shift in gut microbiota in older adults: a randomized, double-blind, placebo-controlled, crossover study. *J Am Coll Nutr.* 2015;34(6):459–469; Miyoshi M, Ogawa A, Higurashi S, Kadooka Y. Anti-obesity effect of *Lactobacillus gasseri* SBT2055 accompanied by inhibition of pro-inflammatory gene expres-

sion in the visceral adipose tissue in diet-induced obese mice. *Eur J Nutr.* 2014;53(2):599–606.

274 ***Lactobacillus helveticus*** Luo J, Wang T, Liang S, et al. Ingestion of Lactobacillus strain reduces anxiety and improves cognitive function in the hyperammonemia rat. *Sci China Life Sci.* 2014;57(3):327–335.

274 ***Lactobacillus johnsonii*** Xin J, Zeng D, Wang H, et al. Preventing non-alcoholic fatty liver disease through *Lactobacillus johnsonii* BS15 by attenuating inflammation and mitochondrial injury and improving gut environment in obese mice. *Appl Microbiol Biotechnol.* 2014;98(15):6817–6829.

274 ***Lactobacillus kefiranofaciens*** Hong WS, Chen YP, Dai TY, et al. Effect of heat-inactivated kefir-isolated *Lactobacillus kefiranofaciens* M1 on preventing an allergic airway response in mice. *J Agric Food Chem.* 2011;59(16):9022–9031.

274 ***Lactobacillus paracasei*** Simeoli R, Mattace Raso G, Lama A, et al. Preventive and therapeutic effects of *Lactobacillus paracasei* B21060-based synbiotic treatment on gut inflammation and barrier integrity in colitic mice. *J Nutr.* 2015;145(6):1202–1210.

274 ***Lactobacillus plantarum*** Hulst M, Gross G, Liu Y, et al. Oral administration of *Lactobacillus plantarum* 299v modulates gene expression in the ileum of pigs: prediction of crosstalk between intestinal immune cells and sub-mucosal adipocytes. *Genes Nutr.* 2015;10(3):461.

274 **respiratory tract priming** Percopo CM, Rice TA, Brenner TA, et al. Immunobiotic Lactobacillus administered post-exposure averts the lethal sequelae of respiratory virus infection. *Antiviral Res.* 2015;121:109–119.

274 ***Lactobacillus reuteri*** İnce G, Gürsoy H, İpçi ŞD. Clinical and biochemical evaluation of lozenges containing *Lactobacillus reuteri* as an adjunct to non-surgical periodontal therapy in chronic periodontitis. *J Periodontol.* 2015;86(6):746–754.

274 ***Lactobacillus rhamnosus*** Cosenza L, Nocerino R, Di Scala C, et al. Bugs for atopy: the *Lactobacillus rhamnosus* GG strategy for food allergy prevention and treatment in children. *Benef Microbes.* 2015;6(2):225–232.

274 ***Pediococcus pentosaceus*** Kawahara M, Nemoto M, Nakata T, et al. Anti-inflammatory properties of fermented soy milk with *Lactococcus lactis* subsp. lactis S-SU2 in murine macrophage RAW264.7 cells and DSS-induced IBD model mice. *Int Immunopharmacol.* 2015;26(2):295–303.

274 ***Saccharomyces boulardii*** Everard A, Matamoros S, Geurts L, et al. *Saccharomyces boulardii* administration changes gut microbiota and reduces hepatic steatosis, low-grade inflammation, and fat mass in obese and type 2 diabetic db/db mice. *MBio.* 2014;5(3):e01011–14. doi:10.1128/mBio.01011-14.

274 **VSL#3 blend** Salim SY, Young PY, Lukowski CM, et al. VSL#3 probiotics provide protection against acute intestinal ischaemia/reperfusion injury. *Benef Microbes.* 2013;4(4):357–365.

274 **by the Mayo Clinic** Mayo Clinic Staff. Mediterranean diet: a heart-healthy eating plan. Healthy lifestyle: nutrition and healthy eating. *Mayo Clinic.* http://www.mayoclinic.org/healthy-lifestyle/nutrition-and-healthy-eating/in-depth/mediterranean-diet/art-20047801. Accessed August 10, 2015.

274 ***Treatment of Abdominal Obesity*** Watson RR, Zuckerman M, Zuckerman E. *Nutrition in the Prevention and Treatment of Abdominal Obesity.* San Diego, CA: Academic Press; 2014.

274–75 *The Elimination Diet* Segersten A, Malterre T. *The Elimination Diet.* New York, NY: Grand Central Life & Style; 2015.

275 *Modern Health Crisis* Prescott S. *Origins: Early-Life Solutions to the Modern Health Crisis.* Perth, Australia: UWA Publishing; 2015.

275 neurological systems Healthy child guide. T*he Neurological Health Foundation* (NHF). https://www.neurologicalhealth.org/. Accessed August 10, 2015.

275 microbial dysbiosis and Schulz MD, Atay C, Heringer J, et al. High-fat-diet -mediated dysbiosis promotes intestinal carcinogenesis independently of obesity. *Nature.* 2014;514(7523):508–512.

275 microbiome and health Jeffery UB, O'Toole PW. Diet-microbiota interactions and their implications for healthy living. *Nutrients.* 2013;5(1):234–252. doi:10.3390/nu5010234.

INDEX

Note: NCDs refers to noncommunicable diseases.